The Hired Lad

About the Author

Ian Campbell Thomson was brought up in Scotland. He was a bothy-dwelling farmworker for nine years, then went to college on a scholarship, followed by work as a farm manager in Surrey. This led to farming a smallholding in Devon and, finally, to Zambia, building a sow unit and growing maize and sunflowers. After retirement, he lived in Oxfordshire. Among his other books are *Mogford's Winning Ways* and *Manager by Appointment*, the sequel to *The Hired Lad*. He died in 2011.

THE HIRED LAD

Ian Campbell Thomson

ORIGIN

This edition published in 2019 by
Origin, an imprint of Birlinn Ltd
West Newington House
10 Newington Road
Edinburgh
EH9 1QS
www.birlinn.co.uk

First published in 1993 by Farming Press Books
First Birlinn Ltd edition published in 2010

ISBN-13: 978-1-912476-70-1

The Author would like to acknowledge the help of
the following: Marie Elder, Drymen Library; Dr
Peter Smith, Treasurer of Drymen and District Local
History Society; John Thomson.

Set in Goudy Old Style at Birlinn

Printed and bound in Great Britain by Clays Ltd, Elcograf S.p.A.

⪜ Contents ⪛

Illustrations

⋲ *Introduction* ⋲

MY story is set in the years immediately following the Second World War, when farm mechanisation, particularly in the south of the country, was gathering momentum. However, many farms in Scotland, where fields were small, the terrain difficult and cash for investment in short supply, were slow to change, and although I chronicle events on just such a small mixed farm in Stirlingshire, my addiction to horses and farm work started much earlier.

Often the only indication my mother had that I was home from school was the sight of my satchel on the path, thrown over the garden gate as I made haste to my adopted farm. Here I might be found driving the cows to their pasture or following the ploughman up and down a field. By the time I was thirteen I could hold the plough and take over for a spell while the ploughman had a break to fill his pipe.

At fifteen I had left school and was doing my best to defer to my mother's wishes that I should try for a white-collar job, but the call of the land was irresistible, and I applied for and was hired in the 'Strong boy able to plough or willing to learn' category, on a farm in Renfrewshire, some distance from my Dumfriesshire home.

Here my room was an attic in the farmhouse; it had a skylight, and from this vantage point I witnessed the bombing of Clydebank. I was alone on this particular night. For a time I watched, scared, shocked and intensely aware of my isolation.

The nearest bombs fell two miles away, and the planes passed over the farmhouse all night. I did what most healthy youngsters would have done: I climbed into bed and fell fast asleep.

The next morning I was ploughing in a lonely field up near the moors, when a stray plane came over flying very low. I heard what I took to be machine-gun fire and something spattered the ground near me. The horses did a 'knees up'. I ran for the shelter of a stone dyke. Hugging this shelter I decided to shelve all plans to become a war hero when I was older. I was unwilling to leave my cover, but my piece-box was at the other end of the field, and I was starving hungry. Needs must when the devil drives.

My employer on this farm used to leave my wages, on a Saturday morning, in the toe of my boot at the foot of the stairs. This was usually some change wrapped in a note and not very bulky. I decided to drop a hint and I pretended to wear my boot all day without noticing the money.

The hint did not cross the brain blood barrier, so I decided to present myself at the hiring fair in Glasgow's horse market, surely about the last of its kind to be held. I walked into the horse market in some trepidation, and discovered something else about myself. I was never going to be a captain of industry, a great negotiator, with an eye for a deal.

I was accosted immediately. 'Are ye fur hiring?'

I was. He offered a wage. I accepted. He gave me half a crown and I was legally contracted. So it was I found myself in Stirlingshire, sharing a bothy with two other lads, also newly hired.

The bothy was really an extension of the stable, comfortless; we slept between coarse blankets; the floor was concrete and uncovered. There was no toilet. The preferred method was to take a spade and find a convenient hedge.

My suitcase rested in other such bothies as I raced along a learning curve, hungry for new experiences, until my story in this volume starts in 1947.

Perhaps I might be forgiven if I romanticise a time when, footloose and fancy free, with all I owned in one old suitcase, I was free come term day—28th May or 28th November—to decide my fate for the next six months. This was the time to negotiate a new contract or go in search of pastures new.

I suppose the passage of time lends enchantment to one's memories. I can close my eyes and imagine the scrape of soil off a plough 'wreast' and smell the sweat drifting back from a fine pair of Clydesdales. I can forget about lacing up hard leather boots with hands painfully chapped with the cold, and the endless days walking behind harrows.

During those early years as an itinerant farm worker, I developed a great interest in the work, and a love of the land, that are with me still. My life has been spent nudging mother nature to produce a little more, to turn sand into gold, but always in the knowledge that we had to work in harmony.

If I draw a give-and-take line through life's good times and bad times, I find a thread which runs through, constant and faithful like a good marriage: a love affair with the land which started long ago.

Who said it wouldn't last?

To the memory of
Donald and Blossom, the horses who shared my work during
the period of this book
and
To my wife Renée,
who joined me in double harness soon after

1
⤜ *Old Blossom* ⤛

I WAS on my way to the match. No, not a football match, not a Rangers-Celtic confrontation, this was a truly rural occasion, the Gartmore and Dalmary ploughing competition.

I wasn't particularly nervous, as I had ploughed the year before, and won a prize too, for the best opening, and was just out of the money for the ploughing. I came fourth in the chill plough class and collected a highly commended card.

That opening was something though. Three up on the land; the first a mere ribbon, the second deeper, the third a bit more, then into the ploughing proper. I remember thinking that such a nice level 'feering' could put a bob or two in my pocket, so I was hardly best pleased with a gallon of linseed oil for the horses.

These and other thoughts passed through my mind as my entourage clip-clopped along the empty road. We had left the farm at six, and it was quite dark, a stable lamp swung on the cart axle, and I was equipped with a flashlight to further warn any traffic. Not that I expected much, perhaps a milk lorry heading for an early collection. Tractors had not yet encroached, although a few Standard Fordsons trundled the fields and the lively Ford 'Fergie' was soon to proliferate.

Sitting comfortably on the horses' dinners, I shared the almost new 'Jack' cart with a loaned plough, a veteran of many matches. Between the shafts was a borrowed horse setting a lively pace despite my firm pull on the reins. Donald, a sturdy dark bay horse who made up half of my regular ploughing team, trudged behind the cart. I hoped he would accept the proposed new pairing.

Perhaps I should have listened to the boss. After all he was well into his thirties, with years of experience, and I had only left my teens behind on my last birthday. He had advised against borrowing the horse. But there, I just didn't want to expose old Blossom again.

It wasn't just that Blossom was a bit of a veteran and there was little flesh on her back despite good care and feeding. Nor even that she would stand at the end of a length, head down among her hooves, lower lip pendulous to a greater than normal degree, alarmingly lopsided behind with one leg relaxed. No, it was more to do with that turned-in front hoof.

Last year when a female spectator had taken a long look and muttered 'Poor thing', I had fiercely rounded on her, explaining that Blossom was a Dunure Footprint mare and carried the best blood in the breed.

She had gone on her way unconvinced, and I was left to ponder that the 'Footprint' must have begat Blossom at the end of a hard day. I couldn't quite bring myself to forgive him for her pigeon-toe.

My thoughts had come round to the nub of the matter. I was going for the money prizes. My pecuniary state almost made this obligatory.

Last year's mistakes were clear in my mind. I had ploughed in clean but 'workaday' harness. Couple this with an old, pigeon-toed horse, and I didn't rate a look in the 'Best turned-out' competition. True, a judge did come to stand by Blossom, but he was only getting out of the wind to light his pipe.

This year the boss had borrowed a superb set of finery from a retired performer. Even this elaborate collection of brass and leather would not have been enough to transform the gaunt Blossom. The energetic gelding between the shafts would look superb, and Donald, in his prime at twelve years old, was in fine fettle.

Then there was the best-looking ploughman prize. Last year I had been unprepared for it. I had just finished my opening when a burst of girlish giggling heralded the approach of the young judges. I had little time. I undid a few shirt buttons, knotted my red spotted handkerchief loosely round my neck, pulled my cap over to one side, and allowed the cigarette I was smoking to droop, Bogart-fashion, from the corner of my mouth.

Three teenage girls peered closely. Still suffering from the nervous tension of creating a prize-winning opening, I found it hard to smile. Perhaps instead I scowled. At any rate the judging trio dissolved into hysterical giggles and, arms around, supporting each other, and still shrilling and collapsing with mirth, they staggered off to the next plot. The prize eventually went to a young man who didn't wear a cap, didn't smoke, had the top button of his shirt done up and smiled to order.

This year I had sacrificed a drape jacket I had been keeping for best, my hair was liberally Brylcreemed, I would wear no cap, my cigarettes would stay in my pocket and I had been working on my smile.

Smiling still didn't come easily, but when the judges came round I would think of the prize money. That should do it.

The field was in sight. The borrowed horse had kept up his fast pace despite my tight rein. Donald on the lead rope through a ring on the back of the cart had registered his displeasure by walking the whole way with his head and neck at full stretch.

An official walked up and took details, logging our time of arrival. I looked around; a horse box was coming through the gate; there were no other teams in sight. We had won the prize for being first on the field. Not a bad start.

By the time I'd found my station and got the horses ready, others were arriving. Further along, the 'swing' plough elite were fussing over their ploughs and horses. They were generally older and just a little contemptuous of those of us steering 'Dux' ploughs with a wheel under the beam.

Donald was smaller than his new partner, but he was the 'land' horse, which would give him a six-inch advantage. Both horses looked elegant in their borrowed trappings. The sun, which took a reluctant watery look at the scene, highlighted the well-conditioned coats of the horses, the polished leather and the chains which I had burnished by working to and fro for an hour in a sack of torn-up newspapers.

I had to admit that the boss had gone to town on the gear. He had also helped me get the plough set for this land. On his advice I had asked the blacksmith to put a bit more rake on the coulter. I felt almost conscience-stricken about the times I had grumbled about him behind his back. He arrived in time for the start.

The borrowed horse stood head up, and there was a flare to the nostrils and a watchfulness about the eye that I was not too keen on. Donald stood quietly, but I could tell, despite the acquiescent droop of his head, that he was rigid with hurt and anger. It was none of his doing that Blossom had been left in the stable. What right had this upstart gelding to take her place?

'I don't like the look of that big horse,' said the boss, and went on, rubbing it in, 'I told you it was risky, not having tried him out. You would have been better with Blossom.'

'But the farmer couldn't spare him till last night. Anyway, he said he was the best furrow horse in the district.' I began to retract some of the charitable thoughts I'd been having about my boss. He seemed to have fallen back on grumbling.

'Well, I'd better lead the beggar.'

The boss walked backwards between the horses for an hour. Donald assumed his normal steady gait. The big gelding, head up, was progressing in short stamping steps. His big feet clipped the furrow and the swingletrees were seldom level.

After an hour the boss said his corns were hurting and he stomped off. I could tell he was not pleased with me.

I took stock of my position and came to a decision. I moved the pin at the head of the plough to give a wider furrow, asked Donald for a bit more pace and had my quarter acre turned over by midday.

My 'feering' was lumpy, the ploughing didn't 'sit up' and my 'hinton' was too deep. The girls had not been round to see my Brylcreemed hair and my buttoned-up shirt. The 'turnouts' were to be judged at the end.

I put Donald in the shafts, and with the borrowed horse on a lead rope behind, still full of energy and pushing up on the tailboard, I headed for the gate. As I left the field an official noted the time and my name. My second prize of the day was to be 'First finished in workmanlike manner'. I then set off for home, making a detour to deliver the out-of-favour horse to his owner.

There was an exchange of whinnies as we entered the stable. I let Blossom out for a drink. As she plodded across the courtyard to the trough, I thought, 'the old lady doesn't look bad for her age.' As always on the farm, I hardly noticed her deformed front foot. I wondered whether, had I listened to the boss, Donald and Blossom might have got me among the money prizes, perhaps even a cup. As it was it would most likely be linseed oil again.

Of one thing I was sure. Whatever it was, old Blossom would get her share.

2
~ *Bloomers* ~

I HAD been staring at the washing line for some minutes when the boss came up behind me.

'It's not supposed to be respectable for a lad your age to be so interested in bloomers,' he opined.

They're so big,' I said. 'Barrage balloons.'

'Wind socks,' the boss replied. Yes, perhaps he was right, twin pillows of grey, ballooning in the wind.

'You know,' said the boss in a reminiscent tone, 'when I was young like you, I once stuffed a pair of those full of hay. "Coup cairts" we called them.'

'Coup cairts?'

'That's right. They had a buttoned-up flap at the back; you could drop it like the tail door on a box cart. And big... every bit as big as Aunt Kit's.'

'What happened...?' I started to ask, but he was walking away, whistling tunelessly.

He stopped. 'You'll need to tie the bottoms,' he advised and dropped a piece of twine on the ground.

It was tempting. They would look even bigger packed with hay. Aunt Kit would have a fright when she came out for the washing.

I liked her, and although I was the hired lad and was not related, I had quickly fallen into the way of calling her Aunt Kit.

I think she liked me; I always got the extra 'tattie' or the left-over slice of meat. She was one of the few people who seemed to appreciate that a young man doing heavy work needed lots of food. She had looked after the boss ever since his parents died within weeks of each other, ten years previously.

But no; I was on my way to the stackyard. I put temptation—at least for the time being—behind me.

It had been a difficult harvest, with much setting up of wet stooks, but finally it was in and we had finished 'theeking' the stacks yesterday. I was about to run a pair of sheep shears round the edges to make the thatch look tidy.

It was on the way back that I succumbed. There was a haystack nearby with a wedge already started. The hayknife and a fork were an open invitation.

As I stood back to admire the magnificence of my sculpture I was assailed by misgivings. Aunt Kit was not as young as she once was; perhaps she would not be amused. There was an amazing amount of hay in there. Enough to feed Donald and Blossom for a few days.

Just for once I couldn't find the boss to ask his advice.

I spent the rest of the day marking out a stubble field for ploughing. A scrape to mark the headland, and openings every forty yards; the sort of work I enjoyed, just me and the horses, the following seagulls and the sweet smell of the turning earth. Gigantic bloomers stuffed with hay never crossed my mind.

I didn't see the boss again until the end of the day. I heard his unmusical whistling as he walked up from the byre. I had just finished in the stable, and our paths converged near the kitchen door.

'One thing I forgot to tell you,' he said in conspiratorial tones, 'those "coup cairts" I stuffed with hay, that was on a neighbour's farm. Nobody ever found out. But of course,' he went on cheerfully, 'I keep telling you about keeping your own doorstep clean.' With that, still whistling, he went indoors.

I knew right away I was in trouble. Aunt Kit's face was like marble with frost on it. I seemed to be on half-rations, but stifled my complaint just in time. The boss was offered the extras, which he accepted, graciously smiling.

The cold war lasted for a week. I felt sure I was becoming gaunt and thin. I took to pulling the odd turnip to ease my hunger pangs.

The boss was helpful through my period of dishonour, explaining that bits of hay and seeds were difficult to remove from undergarments.

I suggested that Aunt Kit might try wearing her 'breeks' inside out.

The boss professed to be shocked at such an indelicate suggestion from one so young, and pointed out the potential problems with inside-out buttons.

Aunt Kit, in the meantime, spent her spare moments picking hay.

At last it was over; the bloomers bloomed once more on the line. Aunt Kit smiled again and became her usual kind self. Extras came my way, I had done my penance.

———

The boss wasn't whistling much those days. He spent a lot of time grumbling about the weather, the cost of a set of shoes for the horses, the machinations of the Agricultural Wages Board, the deficiencies of the NFU, and a whole list of other miscreants all conspiring to reduce him to pauper status.

The toneless whistling, I had worked out, was his way of preventing his face from revealing inner mirth.

Inner mirth was what he needed just then. I wished I could help, but I was walking a careful line. You could say I didn't want to make a bloomer.

Two weeks later Aunt Kit went down with flu and the boss recruited a young girl from the village to help out.

There was a maid's room off the kitchen and she was to live in. Her corn-coloured hair and soft curves drew me like a magnet, but I was out of luck.

Her social life was apparently well booked up. At six-thirty each evening there would be a bicycle leaning near the kitchen door and a young man shuffling and twisting his cap in the back kitchen, waiting to escort her to a function in the village.

Mind you, it was not always the same young man. The one I resented most was the freckled chap who revved a spluttering Francis-Barnett two-stroke up the farm road. In his case I had a secondary grudge; it would be at least two years at my rate of saving before I could contemplate such an outward display of affluence. I would have to consider my options come term time. Perhaps on a bigger place I could increase my experience and improve my finances.

For the present I had enough to think about with amorous youths encroaching on my preserve. I think the boss had an inkling of my unease, because he trotted out his favourite saying about clever dogs not messing on their own doorstep. I think he based that one on one large male dog who seemed to understand this logic.

Top, our bearded collie, was always there to bring in the cows and attended in the kitchen promptly at six for his supper; indeed his farmyard credentials were impeccable. Somehow, though, in between his regular chores he found time to visit the most remote homesteads and the number of crossbred bearded collies in the district represented a few soiled doorsteps.

I was on my way down to the stackyard and my thoughts were pleasantly on the imminent arrival of the Irish 'tattie howkers': the little old ladies with their blackened clay pipes, the barking foremen and, of course, the young female element. There was that little dark one Mary last year, and the big-fronted Sobina—she seemed to like me... And then I was passing the washing line.

It always struck me as funny that every time I stopped to admire the view the boss appeared. He stood at my elbow. Aunt Kit's greys were gently filling like gathering cumulus but they were not the focus of our attention.

'You'd have a job to fill those with hay,' he suggested. 'They're what they call French knickers.'

It struck me that these days the boss was always telling me things I already knew about. Slacken the stack props to let the stacks settle, we don't want the ends of the sheaves pushed up to let the rain in. He had told me that last year, and everybody knew about French knickers these days.

As I plodded off to loosen the props I looked back at the dainty garments dancing vivaciously in the light breeze. I felt as if I was being given a privileged view, rather like an invited guest at a first night. But then I thought sourly of the many suitors and realised that exclusivity was extremely unlikely.

I could hear the boss whistling somewhere. I was glad his inner mirth had returned.

3
⪦ A Bit of Irish ⪧

T HEY spilled over the tailboard of the potato merchant's
lorry and straggled into the field. They were the usual mix
of sexes and ages. Tom Murphy was once again the 'ganger',
and I was glad to see among the colleens the sturdy figure of
Sobina. She was a year older, ripening rapidly into splendid
womanhood.

'Hello, 'Bina,' I called, going up to her. I was a year older
myself and just a mite bolder.

'Hello yourself,' she replied and made a grab for my trousers.
I was unprepared for this amount of friendliness and retreated
in some confusion.

From a safe distance I counted heads. There seemed to be
twenty-five, a fair squad, and I recognised many from the year
before.

Three old ladies were girding themselves with hessian sack
aprons and getting their clays filled and drawing well. One
held out a blackened stem. 'Would you like a wee draw, sonny?'
she enquired.

I declined with thanks and went to consult with Tom
Murphy about the work.

'You'll be carting for us same as last year,' he supposed,
offering me a cigarette.

I nodded. 'Me and Donald.'

Donald stood, resting quietly between the shafts of the
low loader, apparently oblivious to the hubbub around him.

Tom, a veteran of many 'howks', drew deeply on his cigarette, clapped his hands and said briskly. 'Right lad, we'd best get started. We'll have the tatties stacked by the gate; the lorry will be here after dinner to load up. Now if you put the bags round I'll get the folk sorted.'

'Tom,' I said desperately. 'Can you keep 'Bina off my back? She's a bit too much for me.'

Tom laughed. 'She's a bit too much for most folk these days,' he said. 'She'll have your virginity like a shot if you don't look out. Look,' he pointed. 'See that great brute of a chap over yonder. That's her boyfriend, or maybe it's her ex-boyfriend, she's been trying to get rid of him for a while now.'

'He's big,' I said and made a mental note about not giving offence in that direction.

'And the wee shit beside him, that's his mate. A nasty pair taken together, Sobina would be well shot of them. I think she's looking for a white knight to ride to her rescue.'

'White knights live in England,' I reminded him. 'I don't think I'm her man, not yet, and I don't like the look of that pair of villains. Have a word with her, Tom, I'll buy you a pint.'

At the end of the day Sobina came to help me pick up the last of the sacks, heaving them on board effortlessly. On the far side of Donald, she gave me a cuddle. We were not within the boyfriend's vision and she did not on this occasion go for the trousers. Her arms were surprisingly gentle and her body soft and yielding against mine. By the time her lips pressed down on mine I was quite helpless and nearly swooning with pleasure.

The spell was broken with the arrival of the merchant's lorry.

The driver was leaning on the horn. People were scurrying hither and thither collecting their belongings and helpings of potatoes to boil up for supper. The potato merchant's hutted accommodation was six miles away. I was safe from my temptress until the morning. A bit of a kiss out of sight was all right, but it was never worth getting the wrong side of those two toughs... or was it?

I debated this on the ride home in the cart. It was easy to recapture the feel of Sobina's arms and the soft caress of her lips, and it was equally easy to picture those two men, one of them at least enflamed with jealousy, fists and perhaps heavy boots flying. Both lines of thought made me break out in a sweat.

I guessed the field would take a week. We were digging second earlies, sold as a growing crop to the merchant. The amount of 'shaw' still on the crop dictated that they be dug with forks. Yes, I thought, a week to keep clear of trouble. The trouble was, though—'Bina was not the sort to wait to be asked; she was more inclined to grab what took her fancy.

The days ahead were fraught with peril.

In fact I led a charmed life. Any philandering that went on was confined to the end of the day behind the sheltering bulk of Donald's haunches. I remained undetected and began to question the one about not being able to have your cake and eat it.

I felt so good about things that after a few days I decided to meet a few mates in a pub. I felt I had earned a pint and anyway I had a need to reveal to someone my new life of love and intrigue. Perhaps they wouldn't believe me; I could hardly believe it myself. After supper I got my bike out.

My pint lasted an hour, I had been so busy talking. Some of the lads were frankly sceptical, some of the younger ones impressed.

'She can't be that sexy,' said one of the doubters.

I had just pronounced that she was the sexiest girl for miles, when a pair of soft arms enfolded me from behind. On the end of one arm was a pint which she placed before me. 'A pint for my man,' she said in a husky temptress voice. 'And who's this sexy bit of stuff you're all lusting after? I'm real jealous, so I am.'

I puffed up with pleasure and then almost immediately my sails emptied and hung loose. I had a nasty feeling that retribution in the shape of two roughnecks must be no further away than the bar. I looked round nervously.

'They've gone for a piddle,' whispered Sobina, her mouth on my ear.

With that she nuzzled my neck, said, 'I'll see you tomorrow, lover,' and left me.

I had felt the pressure of teeth; I would have to wear my polo neck in the field tomorrow.

I rose to my feet, and with a show of nonchalance took my leave. I collected my bike, mounted and rode off somewhat unsteadily down the road.

Behind in the pub a group of young farm workers sat, stunned and open-mouthed for a time, then as a man carried their empty glasses to the bar. This called for another drink.

The field would be finished by Friday. There was a dance in a local village on Friday night. All my mates would be there and I understood the Irish would be turning out in force. This would include the two I was having bad dreams about, but 'Bina would be there; there would be moments, and she was being the soul of discretion. I could hardly wait.

I dressed with care. Aunt Kit had pressed my trousers. I scrubbed my nails for ages and adjusted the furrow through my Brylcreemed hair. The gold-plated watch my brother had given me no longer worked, but I wore it just the same. I was not completely sure about the tie, and my cuffs kept disappearing up my sleeves if I bent my arms. I would pull them down just before I went in and keep my arms by my sides as long as possible.

Finally, the cycle clips and a running leap into the saddle and I was on my way.

I looked into the hall. Cigarette smoke hung and accordion music blared. The hall was already filling up. Sobina was on the dance floor, firmly clasped in the arms of my arch rival.

I withdrew and repaired to the pub, where I had a double whisky followed by three 'wee heavies'. This lightened my head as well as my pocket, and my return to the hall was achieved with more panache.

I walked straight up to Sobina and whisked her away from under the big man's nose. I suppose I was a reasonable dancer; in any case if my feet got in a 'fankle', Sabina simply picked me up, twirled me round a few times and set me down.

I claimed her most of the evening, much to the admiration of my peers. By the time the Dutch courage began to wear off, I noticed that the rival pair had sloped off.

The evening passed all too quickly, and the band struck up 'Who's Taking You Home Tonight?' 'It looks like you, lover,' said Sobina in a tone loud enough for one of my friends, standing nearby, to hear.

He had just been turned down by his partner and he shot me an envious glance.

We set off; I wheeled my bike with one hand and the other was clasped in 'Bina's. My feet scarcely touched the ground until, a mile or so on, the 'wee heavies' began to trouble my bladder again.

We were passing a derelict roadside barn at the time. 'I'll pop behind the barn,' I said handing the bike over. I trotted off. There seemed to be little relationship between the volume of the 'heavies' and the amount I had passed during the evening.

Relieved, I was about to rejoin my companion when an arm shot round my neck and pulled painfully tight. It was a big strong hairy arm, and I knew it belonged to a big strong rough man.

I was pulled, heels dragging, into the barn, and was pushed against a wall. With his hands now on my lapels the great bully began to explain in horrible detail what he proposed to do to me. In the background I saw the smaller man dancing about making aggressive motions with his hands and feet.

I suddenly realised that I was not afraid, just very angry. I didn't go to the cinema three times a week without seeing people get out of just such a situation.

I swung my arms round and brought them down sharply on his to loosen his hold; at the same time I headbutted him in the face and kicked his shins.

A call for help brought Sobina to the scene. She waded into the big man, taking over from where I had left off. Meantime the smaller man, catching me unawares, kicked me in a spot which was to alter my aspirations for the rest of the evening.

Doubled up on the floor, I watched 'Bina rout out the two of them. She knelt by me. 'You spoiled his good looks for him sure enough,' she said admiringly.

'Some folk call it a Glasgow kiss,' I said weakly. With that she kissed me. 'I'll get you home,' she said. She wheeled me home on my bike. I said goodbye from a jack-knifed position.

'A pity,' she said. 'I thought we might have had a bit of fun... you know.'

'Some other time perhaps,' I said formally. My 'Quasimodo' stance made kissing difficult. Sobina dropped one on the back of my neck and left.

It took three days before I assumed a normal *Homo sapiens* posture. The boss was far from happy. He seemed to have forgotten that Tom Murphy and I had handled the contract spuds without any help from him.

In the meantime my contribution to the work was below par. There were some things I just couldn't do. The boss had to lift the horse collars down from their brackets. This and other functions he undertook with bad grace. He sent me ploughing because he reckoned my new shape was best suited to this task.

He also presumed to lecture me on women, a subject I thought he knew little about.

'In the first place,' he said, 'the girl is older than you, bigger than you, she might have suffocated you.' I knew he meant this literally and was inclined to agree.

'What you need,' he went on, 'is a girl a bit younger, maybe a bit shy, and less experienced than you.' He added as an afterthought, 'If that's possible.'

Nettled by this last remark, I said, 'Why don't you take your own advice?' I was unprepared for his reply.

'Maybe I have,' he said. 'Maybe I have.'

There was a funny look in his eye and his lips were already beginning to purse up for a whistle. Before I could pursue the matter further he turned and walked rapidly away.

———

I was dressed up in my very best and on my way to Glasgow. People, I was sure, would think I was a bank clerk or perhaps an off-duty policeman.

The bus pulled up in Balfron. I noticed a knot of people on the opposite pavement. They looked like the Irish 'tattie folk'; they were, and in the middle, holding court, was the irrepressible 'Bina.

Perhaps if I sat very still she wouldn't notice me; it would only be for a minute. But no—my luck was out, the driver was climbing down. I watched him light a cigarette. He must be ahead of schedule, several minutes perhaps, anyway he was set for a smoke.

There was a tap on the window. I pretended not to notice, but couldn't ignore the second rap which came with a shouted greeting. 'Come on you old bugger, it's me, 'Bina.'

I looked at her through the plate glass. As usual she was bursting out of her blouse.

'We're working at the MacDonald's place,' she bellowed. 'A good crop of "tatties", better than yours. I mean the ones in the field, not the two...'

I held up my hands beseeching her not to go on. People on the bus were turning round.

'Bina was grinning widely. 'I see, a bit sensitive about our tatties are we? A sore point maybe?'

I could sense the rise in interest level within the bus. 'Bye then,' I tried.

'You're going,' she said. 'And the driver not even in.'

I managed a feeble grin.

She went on remorselessly. 'I enjoyed the dance, didn't you? And have you recovered from that bit of fun we had in the barn on the way home?'

People were giving me strange looks. I could sense general disappointment that the driver was climbing back in. The engine picked up, we were moving. I was sweating like a plough horse.

Sobina, keeping pace for a few yards, planted a generous kiss on the glass. 'See you next year,' she called.

I raised a hand and smiled weakly. Perhaps you won't, I said to myself; not if I see you first.

4
⮚ *Staying On* ⮘

W ITH the potatoes out of the way I had the midden to empty. The dung was going on a stubble field which would be the scene of next year's Irish activity.

Endlessly forking dung into the cart was a tiring and tedious job. My attitude to the work was not helped by the boss leaning on the wall from time to time as he passed about his byreman's duties.

Sometimes I would leave a spare graip invitingly near the wall, but the hint was never taken. Such division of labour, I felt, was hardly necessary on a small place, although I would have been hard put to it to make a decent list of the times I had helped out in the byre or cut kale for the cows.

My resentment grew with my fatigue and the seeming endlessness of the task, especially as the boss gleefully wheeled out daily barrowloads to add to the 'mountain'.

I was using Blossom as a trace horse. Not that Donald couldn't pull the loads up the bit of a hill, but more to give the old girl some exercise. Anyway, it made the job more spectacular and helped to relieve the tedium.

After three days of steady endeavour I was making a dent in the heap.

A rattle of milk cans announced that the boss would soon pass with a trolley-full of empties retrieved from the stand. He stopped. As so often happened, I was leaning on a shaft having a quick puff.

The boss didn't smoke and was sometimes at pains to explain that his main objection to smoking was the man-hours (an expression he had recently acquired) lost to the nation by indulgence in the habit.

To back up this theory he would multiply up my lost time by the population of Glasgow, or the country. There was no disputing the case against me, even if the calculations were suspect and much rounded up.

I smoked on, my mood inclined to be defiant. Term time was not far off; I wasn't too sure I would stay on.

The rattle of cans stopped, the boss's substantial forearms rested; there was a far-away look in his eyes; he appeared not to be calculating lost man-hours.

Observing him not observing me, the penny dropped. A woman. He was either in love or contemplating matrimony, or perhaps both. That remark, when I had advised him to take his own advice about women. 'Maybe I have,' he had said.

I walked over. With term day in sight I could afford some bravado. 'What's she like?' I tried a jocular shot in the dark.

'Oh,' he said slowly, 'She's got a real touch with the clusters, and nearly as good in the kitchen as Aunt Kit, and a bit of land coming her way one day.' With that he started. 'I didn't mean to say that.'

'Your secret's safe with me.' I was pleased to have startled the truth out of him. He was normally a close one. Aunt Kit had a range of farmyard similes to describe the degree of closeness. I doubted whether she knew about his extra-mural activities.

'Well, I'd better get on,' I said, stubbing out my cigarette. 'If I want to get this done by term time,' I added, dropping the hint.

I lined Blossom in front of Donald, hooked her chains into the slides on the shafts, and with much clucking and hupping set off up the hill. I looked back; the boss was still standing there. If he was in love with this virtuoso of the milking clusters, and, worse, if he intended to surrender his bachelor state, where would this leave Aunt Kit and, indeed, me? Would I be so well favoured at meal times? Would I get my shirts ironed and my trousers pressed?

Taking another scenario: if Aunt Kit remained in her housekeeping role and the paragon applied her deft touch to jobs other than milking, would the boss have need of me?

Although I couldn't see the boss doing the ploughing and all the dung carting and turnip shawing on his own, I felt a considerable unease. My bargaining power coming up to term time was weakening. Still I didn't expect precipitate action; the boss didn't work like that, and making assumptions about ages, I reckoned the bit of land inheritance was a few years off.

Trundling my loads up and down to the field, hawking the dung out into heaps, I had time to think and plan my term time strategy.

After three loads I had convinced myself that there would be no change in the domestic situation in the immediate future. Two more loads and I had decided I would stay if the money was right. By the end of the day I had come to the realisation that I must sow seeds of unease in the boss's mind.

At supper, as I was accepting some odds and ends which Aunt Kit said would only go to waste, I said, 'I don't suppose some of my mates get such good food. Still, I suppose with a bit more in their pocket they can buy fish and chips or the odd pie.'

I saw an alert look come into the boss's eye. 'The food's part of your wages, lad,' he observed cannily.

I didn't respond. The opening shots had been fired; the skirmishing would continue; battle stations had been drawn.

For my part I wanted to stay, but on better terms. After all, the bargain struck on term day would hold for six months. The boss for his part would, I surmised, want to keep me, but at the lowest rate he could get away with.

For a week he appealed to my better nature. He really didn't know how he could manage without me, he had come to rely on me... the way I had been getting through the work... and Aunt Kit was real fond of me.

When I began to respond with remarks about the labourer being worthy of his hire and suchlike, he backed off and had a quiet few days. I began to wonder if I had overplayed my hand; I had a feeling that this was a lull before the storm.

I had just finished bedding the horses and was having a final sweep up when the boss produced his ace.

He walked into the stable. I recognised his companion immediately. Short, thickset, his bandy legs encased in cord breeches and black shiny leggings, a few years older than me, he had ploughed at the last two matches. Both years he had been among the money prizes. This would have been enough reason for me to dislike him, but he was loudmouthed with it.

'I'm just showing this chap the horses,' said the boss by way of minimal introduction.

'Well, there they are,' I said briefly.

The man looked. 'A good stocky wee horse, short in the leg but thick-bodied, wide across the breechin'.'

Donald, it seemed, had passed muster. I felt slightly better disposed towards this unlovely little man who was about to usurp my job.

He moved towards Blossom. He stood looking down at her front feet for a time, then ran a hand down her back. He said nothing but came to take the boss a few yards aside. I could overhear his hoarse whisper. 'I couldn't work with an old done thing like that. I expect she'll go to the knacker yard soon. We could maybe have a Ford Fergie.'

The boss stood nodding and rubbing his chin. 'Maybe so, maybe so.'

That 'we' enraged me. Who did he think he was? It was still my job. Blossom had years of work in her given good care. I reached for the stable brush.

The boss hurried his guest away before I could do him a mischief.

The next day the boss made a direct approach. 'I want you to stay, lad.' He named a figure which was a bit less than I had hoped for, but my bargaining power was nil. I had already decided to stay and fight Blossom's corner if I was offered terms.

I thought for a moment. I could hardly surrender unconditionally. 'If there's no more talk of Blossom going to the knacker's yard.'

The boss was already nodding.

I decided to press home my advantage. 'And when she does get too old, promise to retire her to the high moor.'

'I agree.' He handed me half a crown. 'Your arles, lad. I'm glad you're stopping.'

I felt he meant it.

Later, while I was washing my hands in the back kitchen, I heard snatches of a conversation in the kitchen. I heard the boss say, 'Of course I wouldn't have let the old horse go to the knackers, she's got a good five years,' and Aunt Kit saying 'But you've agreed to a bit more pay, I'm glad...'

So far so good; it was the boss's next statement which finally made me realise who had won the war.

'Well,' he said, 'I'm paying the lad a bit more, which he's well due, but the other lad, the one with the bandy legs—he was asking a lot more.'

5
⇝ *Charlie* ⇜

WITH the May term negotiations out of the way, I was settled for at least six more months, and with the light evenings, I tended to venture forth on my bicycle more often.

On three nights a week the village hall doubled as a cinema, and a visiting projectionist showed films, many of which were old and likely to break down. Not that this disturbed people unduly. A ritual chorus of catcalls and wisecracks would greet the flickering images as the film was joined, and audience participation might continue for a time thereafter until the film again cast its spell.

The village also sported an ice-cream parlour which served as a pre-cinema meeting place for the young of the village and its environs. An exuberant half-hour among the village girls and the ice cream sundaes, followed by the uncertainties of the cinema, was not to be missed.

With three evenings a week catered for, I would on occasion spend some of my other free time in the Red Lion at Buchlyvie. Here I might meet friends and enjoy convivial company and a glass of beer.

On one such occasion, Andy, one of the group, suggested I call at his bothy for a billy of tea on the way home. It was here that I met Charlie for the first time, and later big Jake, the farm foreman, who had stayed on behind closed doors at the pub, to 'finish his pint' as he put it.

Andy had filled me in about the farm—two hundred acres, two pair and the odd horse, forty milking cows—but he had not explained about Charlie.

As I had lived in outside bothies before, the dilapidated collection of chairs and iron bedsteads came as no surprise to me. An old man sat crouched over a reluctant fire. 'Bluidy peat's damp,' he said by way of explanation.

'It'll boil a billy,' said my friend. 'And I have some bread.'

Bread, this was better than I had hoped for. Being almost always in a state bordering on hunger, bread and a blackened 'billy' of tea seemed like a feast.

'This is Charlie, he's a patient,' said Andy.

I knew, of course, what he meant. Charlie was from an Institution. He would be 'a bit wanting' but able to do uncomplicated tasks. The Institute would clothe him and supervise his welfare. The farmer would provide food, a place to sleep and perhaps a twist of 'bacca and a few shillings pocket money. Given a reasonable employer, this in my opinion was a far better option than being shut away in some Dickensian place of confinement.

Charlie indicated the green and white scarf which was wrapped around his neck. 'Celtic,' he said. 'The chairman sent it for my birthday.'

'It's right enough,' said Andy. 'He gets letters from the chairman as well. He's been a supporter for a long time now.'

Later, when Jake came in, slightly the worse for wear but capable of speech, he suggested that Charlie show me his letters, which were still in their envelopes. I noted the Glasgow postmark, but there was no Parkhead letter heading, and the quality of the writing left a lot to be desired. They were all in the same vein, thanking Charlie for his loyal support and suggesting that if he should come to a match, to be sure and look in to the directors' box for a whisky.

The green and white scarf was obviously Charlie's most prized possession, but there were other trinkets, a medallion to hang on his watch chain, a notebook and other gift shop products. These items apparently arrived on his birthday and there was always a card at Christmas.

I rode home a little puzzled that night, and determined to find out more when Charlie was not around.

My opportunity came the following week. The pub was quiet. Jake was on his own and had just started his evening's drinking. His large paunch sat on his lap and his powerful fist was wrapped round a pint. His expression was dour, and he looked what he was, a rough, hard-working, heavy-drinking son of the soil, able to take care of himself, not the sort to trifle with.

He was pleased to see me. I bought him a pint.

We talked farming for a time, then I said, 'I'm a bit puzzled about Charlie, those letters, the presents, that scarf he's so fond of, who sends them?'

'Ach well, it keeps the old bugger happy, that's the main thing. Come on, you're falling behind. Not much of a drinking pal, are you?'

Jake was clearly trying to change the subject. I ordered him another pint. 'I can't keep up with you Jake; you've had more practice, but I would like to know about Charlie, he seems a decent old sort.'

'Ach well, he can be a cantankerous old beggar, I have to shout at him sometimes to keep him up to the mark, but I make sure he gets his 'bacca and his few shillings. It's hellish the way some farmers treat their patients, but there, you don't want to hear about all that, he likes his Celtic, you would be too young to remember Jimmy McGrory, and Charlie Napier and wee Johnny Crum.'

I knew I was being sidetracked. I was not going to learn anything from Jake. I admitted full knowledge of the old-time Celtic players, and when Andy came in a little later, I left Jake to his serious drinking and tried my luck with my friend.

'Jake's taking on a full load again,' I suggested.

'He's a heavy drinker sure enough,' said Andy. 'But he's a decent enough chap underneath, shouts a lot, especially at old Charlie, gets in fights sometimes, but he works hard as well, and he's good with animals; just about the best horse-breaker for miles. Another thing,' said Andy as an afterthought. 'He looks after his own. If I got in an argument, here, for instance, he'd be right over, and although he shouts at Charlie, nobody else does.'

This gave me a way in. 'All this Celtic nonsense, who writes the letters? I'm sure it's not anybody at the club, but the letters are posted in Glasgow.'

'It's a long story,' said Andy.

I looked at the clock. 'We have an hour.'

'Well,' Andy began, unwillingly. 'Charlie can read. He reads the papers when he can get hold of them, and a long time ago he started supporting Celtic.'

'I can understand that,' I said. 'Supporting Celtic's one thing, invitations to drinks in the boardroom's another.'

'It had nothing to do with Jake, not in the beginning,' said Andy defensively. 'It started on another place where Charlie was sent first. It was a joke really; some builders doing a job on the farm sent Charlie the scarf on his birthday and had a mock presentation on behalf of the club. Then Charlie, with some help, wrote a letter to thank the chairman, which he gave to the builders to post. Of course they didn't, but Charlie was expecting he might get a reply so they started the letters. And so it went on.'

'And now Jake keeps it up?'

That's right. When Charlie came to us soon after, the letters stopped. Charlie kept looking for the post, he went downhill in health, not eating enough, and miserable with it. Jake wrote a letter and posted it in Glasgow. Mind you, he said in it that, as chairman of an important club, he would be too busy to write very often, maybe no more than twice a year.'

'But what if Charlie ever found out?'

'Not too many people know about it. Those that do, realise that if they don't keep it to themselves, Jake'll pulp them. I shouldn't have told you, even.'

I looked at Jake, huge and red-eyed at the bar. 'Don't worry, I'll never breathe a word.'

We were sitting round the dying embers of a peat fire in the bothy when Jake lurched in. Charlie had been nodding off, alternating his snores and snufflings with short chuckles as some imagined happening amused him. Perhaps he was in the boardroom, sipping whisky with the chairman, or discussing the line-up for the next match with the manager.

Jake dropped a crumpled newspaper by Charlie. 'It's Monday's, it was lying about in the pub, you can read about them. Played like donkeys, one-one at Partick Thistle.'

Charlie stirred and snorted. 'I heard that, you thought I was sleeping, but I heard you, I've a good mind to tell Mr Mayley about you in my next letter. You're daft as a brush, and what's more, you're drunk. And you stink like a brewery. I could complain about you to the boss.'

'Aw, shut up, you daft old bugger,' said Jake, and pulling off his outer clothes he fell on the bed and was soon snoring lustily.

In the stable next door a horse kicked monotonously at its heels. 'A shot of grease,' explained Andy. 'Jake's treating it. He's good with horses.'

Honours even, Charlie had again dropped his head in sleep. I emptied my blackened billy and took my leave.

It was four weeks before I again visited Jake's bothy. I had been busy in other directions. The boss likened my activities to the nocturnal journeyings of Top, the cattle dog. At least you both turn up for work, he would say. Privately I thought I had much to learn from Top. His sort of courtship was less fraught with niceties and convention, and of course his workload next day was that of the privileged compared to mine.

Andy was on holiday the evening I chose to visit the pub, but I had a drink with Jake and decided to look in on old Charlie anyway.

Jake thought this was a good idea. 'He's getting on a bit, and the old bugger's been off-colour lately. A bit on the quiet side for my liking, not answering back; I've been thinking of phoning the Institution.' Jake was quiet for a moment, then said, 'It must be hellish living in one of they great dark buildings. I wouldn't want them to take him away.'

'I'll keep him company for a bit,' I said, and left Jake with his thoughts and a freshly drawn pint.

As I leaned my bike against the bothy I thought I heard strange sounds from within, groans and moans and what sounded like distorted speech. Inside I found Charlie on the bed tossing and turning in delirium. I ran to the house and got the farmer to phone for an ambulance; then, leaving him in charge, I rode quickly to the pub to fetch Jake.

The farmer's wife thought it looked like a stroke, and the ambulance men confirmed the diagnosis. 'Nearly sure it's a stroke, missus,' said one of them.

As they were stretchering him out, Charlie seemed to become more agitated. 'Hold on a minute,' said Jake. 'I know what's bothering him.' He took the green and white scarf down off a nail and gently wrapped it round Charlie's neck. 'There you are you old bugger. I'll come and see you.'

Charlie's agitation diminished and he seemed to be trying to hold out a hand.

'He seems to be all right down one side,' said one of the ambulance men. 'I would think he's paralysed on the other, and his speech has gone. He wants you to take his hand, mister.'

'Aye, och aye, so he does.' Jake took the wavering hand in his and held it for a moment. 'Best get him to hospital, and make sure he gets the best, I'll be in to see he does.'

After a spell in hospital Charlie was taken back to the Institution to recuperate, but had another severe stroke and died.

I got news of the death and went round to see Jake. He was seated on a bench outside the bothy. It was a fine summer evening and Andy was off on business of his own.

'Take a seat,' Jake invited.

'Not off to the pub yet?' I observed, stating the obvious.

'No, later, it's just nice sitting here looking at the view. That's the Lake of Menteith over there.' He pointed.

I looked at the distant lake just discernible by the sparkle of sunlight which burnished its surface. 'It's certainly a peaceful view,' I agreed.

'It must be hellish staying in they places, I'm glad in a way the old chap's got out of it.'

'Well, you still have Andy to shout at,' I tried to lighten the mood.

Jake smiled. 'Oh aye, the trouble is he does what I tell him. The old chap used to cheek me back. I still miss the old bugger sitting there with his green and white scarf.'

We chatted for a time about this and that; then Jake said, 'I have a young horse to break come the winter. If you like you could come and help me. It would be a bit more experience for you.'

I accepted with alacrity, and on the way home fell to thinking about Jake. About forty years old, single, with apparently no woman to share his life and his bed. A heavy drinker, with a paunch to show for it. Burly, strong, hardworking, good with animals and protective of those who came within his orbit.

Jake had a reputation as a horse breaker. I could learn from him. He seemed to have taken me into his circle, perhaps because of my concern for Charlie. I sensed a complex man. There was much below the surface, but for the moment I was content to accept his friendship and his tutelage, and shelter under the carapace of his protection.

6
⤳ To Milk a Cow ⤳

I WAS always fascinated by the way the boss ate his porridge. Aunt Kit would serve this in soup plates, but by the boss's plate she would set a bowl of milk.

There was a rhythm about the way he scooped up a spoonful and sloughed it through the milk on its way to his open mouth.

I was watching him finish off the performance by tipping the dregs left in the milk bowl down his throat, and I was thinking of a dredger I had once seen working on the Clyde, when he put the bowl down with a thump and spoke. 'You should learn to hand-milk,' he said emphatically. 'It'll always stand you in good stead.'

'But you machine-milk,' I protested.

'All except one; she won't let her milk down to the machine.'

'But she kicks.'

'Did you ever see her kick me?'

True, I had seen him milk her and she had seemed quiet enough, but I had heard stories.

'You can strap her legs.' The boss was prepared to be magnanimous.

There was no escape, I had to agree, and before I could stop myself I had blurted out, 'If you milk her without a strap so will I.'

It was too late to retract. He was leaning over to clap me on the shoulder. 'I like a lad with a bit of spirit, and I'm glad you're stopping on. Tomorrow morning, then, after you've done in the stable.'

Three weeks had passed since I had heard that conversation in the kitchen. I was acutely aware that I could have fared better in the bargaining, but that was all settling down in my memory bank to be retrieved only at times of aggravation and dispute. Anyway, I had weightier things on my mind, like the secondhand motorbike I had located on a remote farm, and a sneaking ambition to run at the Aberfoyle Games in the summer.

Secondhand motorbikes were at a premium. This one was priced at twenty-two pounds, and I had been saving for a considerable time now. Aunt Kit was holding the money for me and campaigning my cause with crusader zeal.

'If you have any change in your pocket I had better have it,' she would say, and the boss would weigh in with remarks like, 'If you want the name of a good stockbroker...'

But the fund was coming on. I had visions of going to the Games on my motorbike and presenting a competitor's ticket. It would be tight, but I had extracted a promise from the vendor not to sell to anyone else. The deal had been struck, and for good measure, since he was smaller and younger than me, I had given him to understand that I was not the sort to trifle with and hinted at unpleasant consequences should he let me down.

—

The cow looked like any other cow, quietly eating hay. She looked round as I placed my stool and sat down. The eyes were large, placid and undemanding. She went back to eating hay.

I gripped the pail between my knees and reached for the teats.

Suddenly, viciously and quite deliberately she kicked me and my bucket and stool into the grip and went back to eating hay.

I lay among the cow muck bruised and shaken. The boss seemed to find my predicament hilariously funny and was doubled up, with tears streaming down his face.

Finally he recovered enough to say. 'Well, maybe I'd better milk her this morning.' This seemed to set him off again, or perhaps it was the sight of my dung-plastered overalls.

'You'd better stay and watch me,' he said. 'I'll just get a clean pail.' I could hear his trumpeting laughter echoing round the hollow acoustics of the dairy as he barged about among the empty cans and utensils. He appeared carrying a bucket and stool.

'This stool isn't that dirty,' I suggested.

'I shall need that too,' he said mysteriously.

The laughter had stopped but he was still jovial as he walked up beside the cow, slapping her heartily on the rump. She lifted a foot but didn't kick. The boss sat down confidently, the pail firmly between his ample thighs, the spare stool by his right hand. 'Right lass,' he bellowed, reaching for the teats.

The cow stopped eating hay, looked round quickly, then stood eyes front.

I had a feeling that she had flicked a wary glance at the extra stool.

Soon the boss was handing me a foaming pail of rich milk. He stood up and passed his milking stool, then with the other stool still in his possession he advanced to fondle the cow round the head. 'Well done, old girl, we understand each other,' he boomed.

I had the distinct impression there was something deterrent about that extra stool.

'All done by kindness,' he said gravely as we went off for our breakfast. 'You can strap her tomorrow.'

'I will,' I said. 'And I'll have two milking stools.'

We had nearly reached the house door when he turned to me and said seriously, 'A bit of advice.'

'Yes?' I queried.

'Take your overalls off before you go in. You're covered in shite.' And this set him off laughing again.

7
⇽ *Old Top* ⇾

WINTER had the farm in its grip when Top went missing. There had been a fall of snow, with hard frosts to follow. The deer were coming down off the moor to raid the turnip clamps, and none of us was inclined to stray far from the fireside in the evenings. My hands were painfully chapped, and the boss had a chilblain on his heel.

'It's the first time the old devil's missed his breakfast,' said the boss. 'Not that there's any work for him at the moment, but still, it's nice to see the old chap in the yard in the morning.'

Aunt Kit was beside herself with worry. She had collected a few scraps for Top's breakfast and had called him from the kitchen door. Usually he would have emerged from the barn or a shed where he had chosen to bed down after doing his rounds.

It was most unusual. Top's behaviour was normally so predictable. His free time was his own business, but he always attended for his morning snack and his main meal at six in the evening. He would be around the yard during the day in case the boss should want him, and he did most of his visiting in the night or early morning.

Not only was there courting to attend to, but he liked a bit of hunting, digging out rabbit holes and the like.

With the ground rock hard, I didn't think it likely that he could have got himself trapped, but the boss pointed out that fox earths often had big openings.

Worriedly he gave his opinion. 'He could have got far enough down one of those great caves to get digging and get himself stuck. The silly old beggar just doesn't know when to stop.'

When Top didn't appear for his six o'clock supper, I volunteered to go and check a few fox earths I knew of.

The boss got on the phone to ring round his neighbours. 'I would come with you, but my damn chilblain hurts when I walk,' he grumbled.

'Wrap up well; have you no gloves?' asked Aunt Kit.

I admitted that I didn't own such luxuries, but suggested I could keep my hands in my pockets.

She tut-tutted. 'Wait here,' she said and left the room. She reappeared with a hand-knitted balaclava and a pair of the boss's hand-knitted woollen socks which she pulled over my hands.

'If you see those damned deer on the turnip pits, send them back to the moor,' said the boss.

'Oh poor things, they'll be hungry.' Aunt Kit was much concerned about empty stomachs during this hard weather.

'It's not what they eat, it's the mess they make. A bite here, a bite there, and straw all over the place.'

'Maybe if I took a cartload up the moor, it would keep them happy,' I suggested.

'Aye well, and maybe they would walk right past it. I hope that old dog isn't caught in a trap or a snare.'

'I'll get off. I'll look along the hedges for snares,' I said, struggling into my ex-army greatcoat and Aunt Kit's comforts.

My greatcoat had been dyed navy blue and Aunt Kit's comforts were of the same hue.

'You look like the Black Knight,' grunted the boss. 'Get your boots on and get off. Don't get yourself lost.'

The moon was bright, and a full complement of stars sparkled. It was nearly as light as some of the days had been recently, and it was bone-chilling cold. A bit of a wind drove snow particles, and my face felt raw.

I walked briskly in the direction of the moor. The upward gradient stirred my circulation, and, warmed, I could look back and appreciate the beauty of the scene.

About half a mile distant now, the farmhouse twinkled warmth and comfort from its windows, a curl of smoke from the chimney was clearly visible, and the buildings were clean cut against the snow landscape. This was where Top belonged, but he was out here somewhere, cold and hungry, or perhaps even dead.

I pressed on. There was no sign of deer. They would come down in the early morning. I checked a couple of fox earths, then crossed to the neighbour's farm and walked the hedges.

I knew there were no traps or snares on our farm, but had no way of knowing about others. I felt sure that by now the boss would have telephoned all the neighbours, so I was not too worried about being accosted by owners.

I cleared the neighbouring farms and came to a small area of fenced woodland. This might be more hopeful. I climbed the well-maintained fence and entered the relative gloom within.

The trees cast shadows, and the place had a creepy feel after the bright openness of the fields. I found a fox earth and had just gone on hands and knees to listen for whimpers, when a voice crackled out just behind me. 'I have a gun, just get up nice and slow.'

My scalp prickled and shivers ran wild beneath my greatcoat. I got up slowly, and turned to confront a stout man in gamekeeper's tweeds. He had a gun, which, I noted with relief, was pointing earthwards.

'What the hell are you doing looking down fox holes?' the man wanted to know.

'I'm looking for a dog,' I said stupidly.

'A dog!' the man exclaimed. 'Foxes live in fox holes, not dogs.'

'No, no let me explain. It's old Top...'

The man listened patiently as I told my tale.

'I know the dog,' he said when I had finished. 'A hairy old chap, a right old farm dog, a bit of a roamer. By the way, do you know you are trespassing? I spotted you mooching along the hedges when I went out to see to my dogs. I thought you were up to no good.'

'Well, I have to find the old chap, fair means or foul,' I said. 'I'm sorry about the trespassing.'

The gamekeeper laughed. 'Well, maybe another time if you didn't look so much like a commando—you fair put the wind up me.'

'We're about evens then. I nearly dived down the fox hole when you spoke. But you'll keep a look out for the dog?'

'I will that,' said the gamekeeper. 'And I'll ask around.' The next evening I widened the search, and on the third day the boss said, 'I think we have to say Top's a gonner. Nothing could survive long out there in this weather, with no food.'

Aunt Kit was distraught, and I could tell the boss was upset by the gruffness of his manner. Between us we had alerted practically everyone in the district: police, gamekeepers, farmers, my friends, roadmen, railway workers. It seemed that Top had disappeared off the face of the earth.

In the afternoon I decided to replenish the small hay store by the stable. I backed a cart up to a hay stack in the stack yard, and as I climbed down I thought I heard a whimper.

I listened, and again, faint but unmistakable, the whimper of an animal in distress.

I found Top on the lee side of one of the haystacks, sheltered by the overhang, and half covered with loose hay. He had obviously used his last bit of energy to scratch his way as far into the shelter as he could.

He was desperately thin and weak and there was blood on him. I ran for the boss and we split a hessian sack to carry him to the house. We laid him close to the kitchen stove while Aunt Kit warmed some milk.

Top was too weak to sit up, but attempted to lap some milk with Aunt Kit supporting his head.

The boss ran his hands gently over him. 'He's been shot,' he pronounced. 'But he was lucky, nearly out of range, nothing very deep, he'll be all right.'

'But shouldn't we get the vet?' I wondered.

'No, no, he's exhausted, but all he needs is warmth and nourishment. Aunt Kit will see to that. Just good nursing, that's all. You'll be all right, won't you?' There was the slightest movement of his tail.

'But the pellets?'

'The pellets will work their way out and the dog's tongue will do the rest. He'll start licking his wounds in a day or two; then you'll know he's on the mend.'

'But who would have shot him? I know he must be a bit of a nuisance, but a bucket of water would have been enough.'

'It's probably just as well we'll never know. It makes for bad neighbours, and I might feel like putting some shot into whoever it was.'

'I wonder how long he was behind the haystack?'

'Oh, he probably got back the first night. Crawling on

his belly, I wouldn't wonder, by the state of that fore leg. Just couldn't manage the last wee bit.'

In three days Top was sitting up and licking himself. After another well-nourished week he had removed himself to the barn, and a few days later he resumed his night patrols. I was amazed at his powers of recovery, and said as much to the boss.

'Well, the shot was mostly just in the skin. If it had been deeper the vet would have had to pick it all out. It must have been a slow job dragging himself home; he was well peppered.'

'Well,' I said philosophically, 'that's what you get when you hang around places where you're not welcome.'

The boss grinned. 'If you come home with pellets in your arse, I'll get Aunt Kit to pick them out.'

'Would I get a bed by the fire and extra food?' I asked.

'Och no,' said the boss. 'That's just for dogs; you have your work. I'd make sure you had a good injection, though.'

'Why so?'

'Well, you wouldn't be able to lick yourself, not your bum anyway.'

I looked out the window. Old Top was walking across the yard, gaunt and hairy, still a little stilted in his movements.

'I don't suppose Top finds it that easy at the moment,' I said.

With that, right on cue, Top sat, suddenly, in the middle of the yard, raised a hind leg, and swooped, licking industriously.

'Fully recovered,' said the boss looking over my shoulder. 'He can do everything now. I suppose that's what they mean by a dog's life. Never mind, we have our work.'

'Yes,' I said. 'There's always the work.'

8
Getting in Shape

*T*HE spring brought a fall of snow in March. We still had lea to plough and beans to skim in for a crop of mashlum. There was also a surprise announcement from the boss.

'I was thinking of buying another mare,' he said between scoops of porridge. 'She could breed a few foals and help out until Blossom's ready to retire.'

I couldn't conceal my excitement and let out a whoop that made Aunt Kit jump. 'That's great,' I said. 'We could be showing at Stirling Foal Show one day.'

'And in time have a two-year-old for the Lanark Sales,' said the boss, piling it on.

It was more than two months to the May term, and already he had played an ace. But there was more. 'Of course you'll be building the hay stacks this year. The neighbour's chap is getting a bit past it and worried about heights.'

If this wasn't an ace it was a picture card. The boss knew I coveted the job, taking great wads of hay off the grab to make those bun-shaped stacks, growing at the eaves, rounding up onion-shaped to a point.

Elevators were in favour but we still used a pole, jointly owned with the farm next door with whom we 'neighboured' at times of peak labour demand.

Both announcements were great news, although I was a bit surprised about the mare. True, she would be useful; I could plough three abreast if the boss could pick up a secondhand two-furrow plough, and there would be something to sell every year once she had started; but the little grey tractors were about now, and the tall Fordson Major was evident on the bigger farms.

There was a lot to look forward to. Aunt Kit had assured me my motorbike fund was healthy despite the purchase of new boots, which had halted the inward trickle for quite a few weeks, and I had been presumptuous enough to decide to train for the Aberfoyle Highland Games. The country's best professionals would be competing in the open events, but I, being a local resident, was eligible to compete in events confined to a given radius.

The snow didn't hang around too long and the work went on apace. The time had come to start getting myself in shape for the games. Feelings of inferiority plagued me. I was desperately self-conscious about revealing not only my plans but my long skinny legs to the boss, and the thought of walking into the competitors' tent among the big names filled me with dread.

There was a five-acre field up near the moor which was totally secluded. I took to sloping off after work to strip down to the shorts I had used at school (now considerably outgrown) don a pair of sandshoes and gallop round the perimeter to the point of exhaustion.

When this became easier, I trained in my new and very heavy boots. I was proud of these boots. The tackets which adorned the soles in rows were so close together that mud had difficulty forcing its way between and, once in, would remain forever. The toes had been designed for shepherds on the assumption that they always walked uphill, such was the curve at the front. They looked enormous on the ends of my long legs, but I persevered, lumbering through the disciplines I had set myself.

I took to running everywhere. The boss was happy to see me running between jobs. The only time he showed concern was when I arrived at the toilet—which was situated near the byre—rather breathless, my boots ringing out a fast tattoo on the concrete.

'Maybe you had better have just toast and a cup of warm water tonight,' he said sympathetically. 'Anyway, in you go before it's too late.'

The weeks passed and I was quite sure my manoeuvrings were undetected. I never went directly to the field but would saunter off in diverse directions, sometimes making quite a considerable circle.

I was caught completely flat-footed when the boss said to Aunt Kit one night after supper, 'We have a ghost in that field up by the moor. It's white with black feet. A cross between a human being and a Clydesdale.'

Aunt Kit, who had leanings towards the supernatural, was prepared to take him seriously, but I knew immediately that my secret was out.

'You've been watching... I only thought... the local race...'

He looked at me sternly. 'You could win that in those seven-league boots you wear. In a pair of spikes you could do well in the open events. I have a pair somewhere... and a stopwatch.'

He produced a large stopwatch from a drawer. 'I'm your trainer from now on. You'll learn to run even-paced laps, and you run clockwise, right-hand in, at this games. You'd better get used to that.'

When he produced the spikes after much rummaging, I asked the question which changed his mood. 'Did you run once yourself?'

He pretended not to hear and got up abruptly. 'I'm off for a bit of a walk,' he said. He collected his walking stick from behind the door and I heard him call for Top. It was left to Aunt Kit to explain. 'It seems he did try the local half-mile years ago, in his father's day. They didn't get on too well at times and the more his father urged him to train... well, he always was a contrary lad. He came last; of course he was never built for running.'

Through the window we watched his progress. A large figure striding heavily, Top following a wary few yards behind. The boss had reversed his stick and, always the farmer, was scattering droppings and scything at emerging weeds.

I turned to Aunt Kit. 'No,' I agreed. 'Hardly built for running.'

The boss had reached the corner where the horses often stood. Here the targets were plentiful, and he was applying himself with skill and vigour.

'No,' I mused. 'Not a runner maybe, but what a shinty player.'

9
⮞ *The Mating of Fanny* ⮜

*I*T was late in the spring when the new addition to the stable arrived with the boss in a cattle lorry.

She was a delightful little mare. A strawberry roan with a creamy mane and tail. Light-boned, and although not clean-legged, not as feathery as most Clydesdales.

'Name's Fanny,' said the boss briefly. 'Comes from a good home.' My intelligence sources had informed me that it was also a home frequently visited these days by my employer.

'She's a real beauty,' I enthused. 'With the right horse we should get some good foals.'

'Just watch for her coming on then,' said the boss.

'But how will I know?' As far as I knew, Blossom had never come in season.

'She'll wink at you,' he said and, handing me the halter shank, walked off whistling.

Nonplussed I led Fanny into the stable, fed her, rubbed her down with a wisp of straw and generally made a fuss of her.

A week later I knew what he meant. I walked into the stable in the morning and she was neighing and her rear end was winking enthusiastically.

When I saw the boss I told him, 'The mare's on; she winked at me.'

'Right,' he said. 'We'd better arrange for her to have a visitor.' I had seen the stallion pass as he walked the district, but never at close quarters. He was a tall black horse with great hairy feet, and led by a thin man with a saturnine cast of features.

At close quarters I realised that compared with my work horses, this was a giant. A long mane hung about his face, his nostrils flared and his eyes showed a lot of white; his huge feet beat a tattoo on the cobbles as he answered Fanny's whinnies.

The boss had only stayed long enough to introduce the stallion leader as 'the groom' and to assure him that I would do all that was necessary.

The groom got off on the wrong foot with me by calling me 'boy'—which I was no longer prepared to tolerate—and referring to Fanny in derogatory terms.

He left me holding the monster, sniffing and intent on getting over the stable door, while he went inside to inspect the condition of his client.

'Boy,' he commanded in a high nasal whine as he again took charge of his animal. 'Take the wee bachle round that corner. She's on well enough. Just stand her there and I'll bring the horse. Oh,' he added unnecessarily, 'keep her arse end this way.'

As I made my way into the stable he was still grumbling. 'I only hope she can stand the weight. She's hardly the sort of Clydesdale we're used to.'

Backing Fanny out of her stall I could still hear his voice, thin and querulous. 'Come on, boy, I can't hold this beast all day.'

In position round the corner of the stable, I waited, nerves tingling. The approach of the stallion was heralded by a tremendous clatter of hooves accompanied by high-pitched words of encouragement from the groom.

I tightened my hold on Fanny's halter and braced myself for the onslaught.

The massive head appeared, much higher than I expected, mane tossing, teeth now visible to add to his terrifying image. He mounted mightily, the lightweight mare lurched forward; I had a glimpse of an enormous hairy hoof arriving not far from my head.

The groom was swearing and making adjustments behind; Fanny stood spread-legged. My nerves were clanging like church bells; and then it was over and I was leading Fanny back to her stall.

I emerged, drained and shaken by the whole traumatic business.

'Boy,' shrilled the groom. 'A bucket of water.' He was preventing the huge member from returning to its sheath, pulling it down for a wash.

'A bucket of water—what?' I said.

'A bucket of bluidy water, now, boy,' he screamed, obviously angered and agitated.

'Get it yourself!' I said and walked off.

I looked back at the tableau I had left behind. The stallion was standing quietly waiting for his wash. The groom stooping to his task was looking after me open-mouthed.

I started back to get the water. I had no quarrel with the splendid great animal who was to be the father of Fanny's foal.

'Say please,' I said to the groom.

He hesitated and I started to walk away. 'Please,' he ground out savagely.

I went into the house to get some warm water. As I was making sure it was just lukewarm I heard his reedy voice. 'For God's sake, boy, where's that bluidy water? My back's killing me and my arm's going numb.'

Instinctively I reached for my cigarettes. Another minute wouldn't hurt. Then I thought of that great proud horse standing there denied his privacy and perhaps feeling cold. 'Just coming,' I called.

I doubted that I had made my point with the groom, but I had at least caused him some discomfort, and with a bit of luck Fanny would be in foal.

I felt a warm little glow of happiness. Nobody could take that away from me.

10
⤳ Conversation ⤶

IT was May and turnip hoeing was going on apace. A spell of dry weather was on our side and the boss and I were spending the bulk of our time at this rhythmic but tedious task. Push, pull, push, pull, scrape, scrape, and the plant we wanted to keep always falling away from the hoe blade... well, almost always.

We were running out of topics of conversation. My training schedule had been dissected and finally put away. The boss talked about the top professionals in terms that might indicate a close personal friendship, which I knew to be false.

Work items had been discussed from all angles. My proposed promotion to haystack builder had been aired, with advice on protocol.

'The man on the stack's in charge,' the boss had been at pains to explain. 'He controls the man on the grab.'

'How?' I had wanted to know.

'By shouting and swearing if needs be. Small grabs round the outside, big ones to heart up.'

Last year the boss had been working the grab. I hoped he would remember our conversation when the time came.

We had been hoeing in silence for some time when I thought of something. 'I like your new hat,' I said.

He looked pleased.

'Reminds me of Edward G. Robinson. He always wears a hat like that... you know... in the films.' I could tell he was about to argue the point.

'More like Anthony Eden. A sort of businessman's hat... or a politicians's.'

I conceded the point, although even in his going-out suit and his black hat, I thought the boss still looked like a film gangster. I mentally ruled out the lead role. He was more like one of the bulky men who lurk in the background uttering the occasional 'yeah, boss', and with a gun in a shoulder holster.

Half an hour of quiet hoeing passed, and then I brought up a problem which had been on my mind for some time. 'That boy, the one who's keeping the bike for me...'

'Yes,' said the boss encouragingly, sensing that a problem was about to provide some mental stimulation.

'Well, he's pushing for the money, says he has another offer.'

The boss was nodding sagely.

'What should I do?'

'Go and bash the shite out of him,' said the boss helpfully.

'Well, it's not as simple as that.'

'Oh, why not, seems simple enough to me.'

'Well,' I said lamely, 'what with me being in strict training and everything, I shouldn't risk getting injured.'

The boss maintained an unhelpful silence.

I went on desperately. 'He's growing like a weed. He's as tall as me now, and broad with it.'

'Well, that's different,' conceded the boss. 'Talk to Aunt Kit.'

The matter was discussed over supper. Aunt Kit got in a rare state about it. 'The rascal,' she stormed. 'You only owe six pounds. He's had his money regular; what does he want, blood?'

'It could come to that,' I conceded miserably.

'Right, we'll pay Shylock a visit. You two get changed. We'll need the car, it's a fair way.' I felt greatly cheered. The boy was certainly behind Aunt Kit in width and didn't have anything near the general bulk of the boss. 'Great,' I said.

'Hold on,' said the boss, obviously less than happy about it. 'I was going out... er... on business.'

'What kind of business?' asked Aunt Kit sharply.

'Oh er... Jessiman... the feed merchant. We have to talk about... er....'

'Go and phone him. Tell him you'll be a bit late.' Aunt Kit was in no mood to quibble.

I heard the boss at the phone in the hall speaking very quietly. He emerged looking as if he had taken a beating.

'Mr Jessiman wasn't best pleased,' I suggested pointedly.

He glowered. 'We'd best get changed, and hurry up about it, I can't spare much time.'

When we met a few minutes later he was wearing his suit, his black hat and an ill-tempered scowl.

As we made what speed we could in the old Vauxhall he grumbled about the distance, the length and condition of the farm road, the rushes growing in some of the fields and the poor state of the fences. By the time we arrived he had worked himself up to a high peak of belligerence. 'I'll deal with the wee beggar,' he said, obviously anxious to short-circuit any negotiations.

'Stay where you are,' snapped Aunt Kit.

The motorbike stood outside the bothy. I went inside to procure the victim for Aunt Kit. They met in a sort of no-man's-land between the car and the bothy. It was no contest. Aunt Kit's bulk dominated, her tongue whiplashed and finally, being a fair-minded person, she produced three pounds of her own money. 'You'll get the other three pounds in two weeks. In the meantime if you renege on us you will deal not only with me but with him.' She pointed dramatically toward where the boss sat in the car, a dark, sinister, black-hatted, bulky figure.

From beneath the black brim of his hat he directed a look of sheer malevolence. This was the boy who had not only ruined his evening but got him a lot of abuse on the phone. His lips curled in a snarl of hate.

Aunt Kit called, 'I've given him three pounds.'

The boss, busy with thoughts of vengeance, only heard 'three pounds'. So Aunt Kit wants me to pay the beggar off, he thought. Just as well, then we can go. He stabbed a hand into his inside pocket, reaching for his wallet.

'Oh my God,' said the boy, 'I'm dealing with the Mafia!' and he streaked off to slam the bothy door shut behind him.

In the field the next day I asked, by way of making conversation, 'How did you get on last night... you know... later?'

A reminiscent look came into the boss's eyes. 'Fine, just fine. It does you good to have a bit of a tiff now and then. It can be quite nice making up.'

'With Mr Jessiman?' I said.

Taken aback he resorted to bluster. 'Ach well, we'd better get on, there's rain coming. Look at those clouds over Buchlyvie.'

He set a pace I was hard-pushed to match. Conversation, I decided, had its drawbacks.

11
⮐ *The Moment of Truth* ⮐

'COMPETITOR, eh?' The car park attendant, I thought, sounded a bit doubtful.

'Local half and open mile,' I said as nonchalantly as I could. Butterflies had already made their presence felt in my tummy and I had come to understand the animal 'evacuate for flight' theory. I had previously thought cows plastered the gangway out of sheer spite when they were let out.

'Is there a toilet?' I asked.

'A latrine, but competitors usually go behind the tent for a run off.'

Sitting astride my recently acquired motorbike, I leaned over to check that the oil drip feed was working. Having little or no mechanical knowledge, I had become preoccupied with the oil, visible through a glass cover, feeding drop by drop into the intestines of the engine. I monitored this as diligently as a surgeon checking a heartbeat.

'Nice old bike,' said the man.

'BSA 1932 model, 250cc, four-stroke, side-valve, drip-feed oil, see for yourself.' The man observed the miracle of oil being pumped by the droplet.

'Will she be safe enough here?' I wanted to know.

'I'll watch out for her. It isn't every day I get a valuable machine to look after.'

'I should give you something, but the truth is I'm skint.'

'Give me a shilling later out of your winnings.' The man

seemed to find something funny in this and was laughing as I left and headed for the competitors' tent, calling at the latrine on the way.

I was standing outside the tent wondering whether to go in or go home when a youth nervously sidled up. 'Is this the competitors' tent?' he asked.

I pointed to the large notice over the door.

'Are you going in?' he persisted.

I nodded.

'I'm a bit nervous,' he confessed. 'Can I come in with you?'

We went in together. A scene of considerable variety met our eyes. Dancers, pipers, throwers, runners. A giant in a kilt was demanding, 'Who the hell's sitting on my throwing shoes?'

'George Clark,' my new friend informed me.

I stripped down to my outgrown football shorts and white sleeveless vest. Across from us a group of runners were resting or rubbing oil on their legs.

'The one in the white strip is Peter Alwell of Kilbirnie'—my companion seemed well informed—'and the older one is Kenny of Wishaw, getting to be a bit of a veteran now but still one to watch, and Flynn of Niddrie, Crawford of Uddingstone...'

I rubbed my legs with the olive oil and wintergreen that my trainer had furnished me with and launched into the stretching exercises he had been insistent about. I was glad Alwell was in white. Most of the others were more colourful in reds, blues, greens, some with satin or silk finish.

My friend was down to shirt and baggy shorts. 'I need a pee,' he said.

I wasn't averse myself. We went behind the tent and emerged as our race was called.

'Soon be over,' said my friend nervously.

The butterflies in my tummy had found a new lease of life. Miserably we trooped to the starting line.

It had never occurred to me that the race would be other than a scratch event or that I would be back marker. The handicapping was arbitrary. Perhaps I owed my place to a passing resemblance to Peter Alwell, favourite to win the open race: white strip, long thin legs, smelling of winter-green, finely tuned, feet dancing.

Freed from my heavy boots, my feet felt feather-light. I looked up the field. My friend in the baggy pants had twenty yards, some schoolboys were ahead on thirty-five. There was a mix of football and rugby jerseys, a few in long trousers, others in full running strip.

The gun went, and I was quickly into my stride, running comfortably, remembering my training geared to the boss's stopwatch, no tense muscles anywhere, conserving energy.

'Don't worry about the others,' the boss had said. 'Run your own race.'

Despite myself I did begin to worry about the others. The race was half gone and nothing had changed. My friend was still scuttling along twenty yards up, the schoolboys were thumping along thirty-five yards ahead, and then, quite suddenly as we came round the bend into the back straight of the final lap, I found myself alongside baggy pants; and the others, including the front marker schoolboys, were coming back towards me.

I ran through the field in the back straight and came round the last bend well clear, to run on and breast the tape, arms extended in triumph. Looking back I was pleased to see my friend in the baggy pants collapsing over the line for second place.

The boss arrived in the afternoon. He had been at pains to explain to me that someone had to stay home to do the work. Since he had missed the race I went through it in detail about fifty times. 'All down to proper training and coaching,' he said. 'We'll see how you do this afternoon.'

The open mile was another story. Off forty-five yards I battled through to the front, only to be overtaken by at least a dozen brightly garbed pros, with Peter Alwell ghosting through in white to run out a clear winner.

Later in the tent I nearly burst with pleasure when Alwell said to me, 'I expect you'll be on the circuit now, Luss, Strathallen, Blackford...' As I was leaving, others called out, 'Well done, son... see you next time... we'll have to watch out for you in a year or two...'

With all this swelling in my head, smelling like a horse that's been rubbed with liniment, feet hardly touching the ground, I went in search of my motorbike.

It started first kick. I leaned over to check the oil drip. It was working; the bike was alive and well.

A voice said, 'I kept an eye on her for you.'

'Thanks,' I said, and was about to ride off when I remembered something. 'I think I owe you a shilling.'

12
⤚ *Big Bertha* ⤙

JUNE was a frustrating month for the boss. He had been unable to demonstrate his normal haymaking strategy.

From about the middle of the month he was anxious to get a few acres cut, but each time we were on the verge of getting the horses out it would rain. Not that it rained a lot, just little and often enough to annoy.

'If you haven't got a bit down when the weather comes,' he was heard to say with increasing frequency as the month progressed, 'you have nothing to get on with.'

Last year it had paid off. We cut a piece during some catchy weather and were able to take advantage of a sudden burst of hot weather to get ahead of our neighbours.

It was the second week of July before a mini-heatwave struck.

The boss had the horses out by five in the morning. He sat on the reaper till I took over at seven. At eleven the horses were back in the stable resting during the heat of the day. The weather held, and this was our daily routine until the cutting was done.

The ripe grass dried quickly and was all up on tripods in just over a week, without a drop of rain on it. It was a smooth and weather-blessed exercise, but the boss felt it could have been even better if he had been able to anticipate the weather and 'had a bit down'.

I was looking forward with mixed feelings to building my first haystack in a few weeks' time. My optimistic vision of perfect symmetry was sometimes troubled by an awareness that stacks could decide to lean and, unlike the much-admired Tower of Pisa, usually ended up supported by a battery of props. Lapses of this sort made good topics for enlightened pub conversation.

However, my youthful optimism didn't entertain such possibilities for long. I was more inclined to see myself high on a perfect stack directing a stream of advice and complaint to the little people down below. I doubted whether my eloquence would match the colourful invective of the retiring stack builder, but he had had years of practice.

My grandiose thoughts must have prevented me from noticing that Aunt Kit had been a bit quiet for a few days, but finally I began to think something was amiss.

'It's Bertha,' she explained when I questioned her. 'The boss is talking of getting rid of her. She's too big, is eating him out of house and home, and she only had six last time.'

'Well, you have to admit he has a point,' I said, thinking of the giant Large White sow, too big for her sty and the mother of so many litters everyone had lost count.

'She's a nice old thing, quite a pet, and I'm fond of her.'

I had to agree. Bertha spent much of her time out of her sty, amiably scavenging the farmyard for scraps; there was nothing she wouldn't eat and nothing edible was ever wasted. Her good nature and extrovert personality had established her as a firm favourite with postmen and other casual callers. She would grunt pleasantly as they scratched her back and lie down and roll over to expose her huge underline for similar treatment.

The cheesecloth-wrapped items hanging from the ceiling hooks were most acceptable contributions from big Bertha, but I could see the boss's point. A sow with half the body bulk to maintain could perhaps produce twice the piglets and leave some profit. I suspected that the boss was also not unaware of the carcase value of his super heavyweight.

The sight of Aunt Kit's despondent face at supper that night prompted me to step tentatively into the arena.

'The news is, you're thinking of selling Bertha,' I tried.

'You shouldn't believe all you read in the papers,' said the boss noncommittally.

I rethought my strategy and decided on a direct approach. 'Are you thinking of selling big Bertha?'

'Thinking, just thinking,' said the boss uncomfortably, avoiding Aunt Kit's eye.

Glum-faced, Aunt Kit went to see to things in the kitchen. Over her shoulder she said, 'Can't think what harm the poor old thing's ever done you.'

With Aunt Kit out of the room, the boss leaned forward and, low-voiced, explained his line of thought. 'The harm is she's had three small litters in a row, and after we've had two pigs out to hang, the rest hardly cover what she eats. Besides,' he added reflectively, 'the market's strong for sows just now, and she must weigh as much as a medium-sized hippopotamus.'

In the knowledge that a good weaner would pay my wages for a week, my next remark was tongue in cheek. 'But surely,' I said, 'to a man in your position the difference between a litter of six and a litter of ten is only a few pounds.'

'Only a few pounds,' he echoed. 'Let me tell you, son, that in business every little counts. Look after the pennies...'

The lecture was drawing to a close when Aunt Kit came in. 'The boss was just telling me how every little counts in business,' I explained. 'I think if Bertha had ten this time he might give her another chance.'

'Well,' the boss demurred, and I knew he was thinking

carcase value. 'That's settled then,' said Aunt Kit quickly. 'If she has ten or more she stays. She's due in a week or two.'

The boss, outmanoeuvred and in any case a little bit in awe of Aunt Kit, muttered agreement.

—

It was time to get the hay bogies out ready for hay leading and to prepare staddles for two round stacks. The harvest pole arrived from our neighbour. This uncomplicated crane arrangement was an effective enough tool for hoisting wads of hay stackwards. It consisted of a tall pole fitted with a swinging jib, a long cable, and a grab to handle the hay. The grab was raised or lowered by a cable attached to a horse which plodded to and fro as required. The jib swung the hay into position. There was the usual fuss and bother of getting the pole upright and securely guyed and pegged, the jib in place, the cable attached to the grab and run down through the eye of the jib through pulleys on the pole to ground level and hooked to the swingletree to which Blossom would be yoked in the morning.

We were standing back admiring our work when I caught sight of Aunt Kit hurrying across the yard in the direction of Bertha's sty. She was carrying a tattie basket which I guessed would contain comforts for new-born piglets.

'The sow must be farrowing,' I said to the boss.

A little later we leaned over the sty to watch proceedings. Bertha was grunting happily and pushing out piglets with the efficiency of a sausage machine. Aunt Kit squatted behind, drying the new arrivals with bits of towel and placing them in the basket where a sprinkle of straw covered a hot-water bottle.

'How many?' asked the boss.

'Ten,' said Aunt Kit. 'No, make that twelve,' she added, as two more followed in quick succession. 'The way she's going she could have twenty.'

The boss reached over and picked up the basket to examine the writhing bodies within. 'I think she might well have twenty. They're ever so wee, I wonder how many will survive.'

Aunt Kit looked him straight in the eye. I had a distinct feeling that the boss was about to be hoist with his own petard.

'It's like you said,' she reminded him. 'Every little counts.'

13
∽ *A Job Well Done* ∽

*I*T appeared that hay leading was to be blessed with fair weather.

The first stack was up to eaves level, and my close inspection could not detect a lean in any direction. Even the deposed stack builder, who came over each day to drive one of the bogies, could find little to crib about. Glad perhaps in one way not to have to perch on the dizzy heights, he seemed a little bit put out that my stack had not fallen over.

Beginner's luck? Or was it somehow connected to the vast amount of nervous energy I had expended worrying beforehand?

The job had fallen into a nice routine. The neighbouring farmer and his employee Harry came over after breakfast. Harry insisted on bringing his own favourite horse, a massive grey Shire. I was a bit put out with him because, with three horses on the place now, we had no need of his Shire.

The boss was vocal and repetitive about the fact that he had thirty-eight cows to milk while the neighbour had only ten, and that he had to get up half an hour earlier to suit the job.

I resisted the temptation to remind him that it was his hay, and contented myself with a grumble about having to trail all the way up to the moor to catch the horses in the morning. As it was I had harnessed them and circled the stack a few times, viewing my work with the critical eye of an architect, by the time the others were ready.

From my elevated position on the stack I could see it all. Out in the field the neighbour, who was working Donald, was winching a pike on to his bogie; the grey Shire was spanking home with one on board, his elderly driver perched on a corner of the bogie.

I wasn't too keen on the way the neighbour jerked on Donald's bit each time he backed him down the stackyard to the grab. Neither, I guessed, was Donald. I could tell by his laid-back ears and his general air of irritation that it was only a matter of time before an arm was nipped.

The neighbour seemed to have decided to give the boss a hard time on the grab, hurrying to keep hay in front of him all the time. The boss, in turn, keeping his end up, sent the hay up in massive quantities with a total lack of response to my pleas from above. The small boy who appeared from somewhere each morning to lead Blossom, the motive power for the hoist, was castigated by all and sundry. Blossom, for her part, plodded backwards and forwards throughout the day, totally oblivious to the frenetic activity around her.

With no helper on the stack I was more than a little dependent on the cooperation of the grab operator. When a massive wodge of hay appeared swinging over my territory, I would guide it with my fork to the selected spot and bellow at the boss to tug the release cord. With one eye on the cavalry charge from the field, and apparently afflicted by temporary deafness, the boss was inclined to pull the cord when he felt so inclined, which was, more often than not, in advance of my call.

By the end of each day my exhaustion was total and my understanding of last year's elder statesman, now happily jogging to and fro behind his great showy horse, was more complete. As the stack grew higher, my lines of communication with *terra firma* became non-existent. Despite the fact that I was now 'heading' and my surface area was much decreased, the hay still arrived in huge portions. I could see the boss, far below, ramming the forks of the grab into the hay with all his considerable strength. For good measure he would add some foot pressure. My pleas for a ladder to be placed handy were ignored.

I had a feeling I might be overwhelmed by one of these large deliveries, my footwork in the restricted space would be inadequate and I would be buried under half a ton of hay. I wondered if anyone would notice.

Thankfully a bogie shed its load in the field. The delay allowed the boss time to procure a ladder and I came down for a look.

With no hay in front of him the boss had time to expand on what was currently his favourite theme. 'It's like I always say,' he declaimed. 'Building a stack is a partnership. The man on the grab has a lot to do with it. It's knowing when to pull the trip cord that counts...'

I walked well clear of the stackyard, counted slowly up to ten and lit a cigarette.

I had to agree that the job had been well orchestrated and, with the weather holding, had been accomplished in record time. The hay had gone into two round stacks, with residues packed into the stable loft and a shed handy for the byre.

I was pleased to see my creations settle nicely and the trials endured in their construction behind me. I was able now to start raking off the loose bits and generally pull and tug to get a good watertight shape.

'If you "pook" much more we'll need to build another stack,' said the boss, observing my efforts.

I spent much of my spare time 'pooking' and shaping up, and when finally the tops were thatched and finished off with a 'toorie', I felt I had created work of considerable artistic merit.

I inveigled Aunt Kit out for a look one evening and I was modestly acknowledging her words of praise when the boss happened along.

He stood rubbing his chin for a moment, then said, 'We did a good job there, lad. It's as I say; it's a partnership—'

I started counting up to ten.

14
⬷ *No Longer a Journeyman* ⬶

BETWEEN hay harvest and corn harvest the boss stepped up his courting activities quite markedly. I knew from the number of evenings he went out wearing the correct regalia.

The black hat was kept more or less permanently in the car. Perhaps he thought that by leaving without it, he presented a more casual image for Aunt Kit's appraisal, although indeed he always contrived to effect his exit when she was not in the house. Once in the car he would don the black hat, using the rearview mirror to check that it sat squarely above the ears and tilted forward to shade the eyes.

It was difficult to reconcile this raffish Romeo with the casual, ambling person who once considered clean overalls and wiped boots adequate for most occasions.

Ostensibly he was attending farmers' meetings, committees and so on, but, not known for his subtlety, he left clues which I soon was able to spot. Although the suit was worn at all times, committee meetings might be attended tieless and wearing brown boots. His exit would be marked by the banging of doors and the sound of his heavy footsteps.

Courting always called for a tie, suede shoes and a quiet and somewhat furtive departure.

On such occasions I might say, 'Not another NFU meeting, boss?' and he would stop uncertainly and perhaps mutter, 'Something like that' or 'Somebody has to do the committee work,' glare at me, but save the telling-off till next morning.

I felt safe in the knowledge that he didn't want me to drop any hints to Aunt Kit, and she in turn seemed content to pretend she didn't know that anything was going on. Perhaps she thought that given time this phase would pass.

So the game of pretend went on. By now I reckoned the boss should have brought his lady home to tea, but as he had not explained his involvement to Aunt Kit in the first place, and she had not indicated any awareness of it, the result was something of an impasse.

Perhaps my interest in the boss's activities was in some measure due to a lack of excitement in my own life. I now derived little aesthetic pleasure from viewing my two haystacks. No longer tinged lightly with the browns and pale greens of new hay, crowned with golden thatch and upstanding 'toories', they stood weathered now to a drab darkness, 'toories' awry, just two functional fodder stores.

Harvest was some weeks away, ploughing even further. The horses were getting fat up on the moor. Most days only Donald was needed. The boss said he had 'punched' a foal in Fanny, but I reckoned that the excellent body and coat condition that goes with motherhood was a more reliable guide. I spent time horse-hoeing, scaling dung and hacking at an overgrown hedge to provide material for stack bottoms.

If the work was presently not challenging, my social life was even less so. I felt disinclined to visit the pub and mingle with the other lads. The girls I met at local dances didn't interest me. I realised that this malaise had afflicted me from the day of that first chance encounter with Penelope. I had been cutting a hedge on the roadside when she came along mounted on a rather tired piebald pony.

'Hello,' I called cheerfully, and the piebald, exhibiting one of Blossom's favourite traits, stopped immediately.

'Hello,' she responded uncertainly. 'You're—'

'Oh, I'm just the lad who does all the work around here, but I know who you are; your name's Penelope.'

She smiled then, but didn't pursue the conversation, and the pony, responding eventually to a succession of heel prods, bore her off.

Her smile and her somewhat fragile beauty haunted me for weeks; she was just what the boss had once suggested for me: younger than me, demure to the point of shyness, petite and, I was sure, suitably inexperienced.

I ventured to mention to Aunt Kit that I had met Penelope and rather liked her. Aunt Kit gave me a concerned look, then said, 'Best leave her be; she's a farmer's daughter.'

At first I didn't fully understand. Then it dawned on me: an obstacle lay between me and farmers' daughters, a barrier of class, and possessions, and money and prospects.

In the most offhand and casual way, I mentioned my meeting with Penelope to the boss one day as we were burning up hedge cuttings.

'Penny's a nice enough lass,' he said. 'A bit lightweight, the nervy sort, hard to handle. She'll marry a farmer's son one day. The fathers will set them up in their own place, you'll see.'

I said I saw, and looking at the converse of his views on Penelope, and taking account of his dismissive tone, I conjured up a mental picture of his preference. Well fleshed, placid and easy to handle, and presumably, like Penelope, destined to marry a farmer.

I now knew where the boss's lady friend lived and that she answered to Peg. My informants were also able to tell me that she helped her father and brothers run a fair-sized farm with a herd of black and whites, some of which (my sources believed in going into details) were Holsteins imported from Canada.

One evening, cruising on my motorbike in the area, I spotted the boss's Vauxhall parked in a gateway. The courting couple sat chatting with space between. I felt that little progress was being made. The boss, if not an apprentice at this courting business, was certainly no more than a journeyman.

For a time my own barren prospects in this department bothered me, as did the thought that social divisions, not apparent when sharing the work in the fields or convivially gathered round the kitchen table for supper, nevertheless existed off the farm.

Another evening, my youthful spirits burdened with heavy thoughts, I was cruising on my motorbike. Pottering slowly along an empty road I spotted a car parked nose first in a gateway. Even if I hadn't known the number I would have recognised the boss's car from the hen droppings on the roof. It appeared to be empty. I throttled down and stopped.

Suddenly a black hat appeared brim high at the rear window. Beneath the brim a pair of eyes glared.

I let in the clutch and throttled away. The boss would have something to say to me in the morning. But credit where credit is due; I had a feeling he was no longer a journeyman.

15
~ *A Long, Hard Harvest* ~

*T*HE ripening corn brought a flurry of activity. The binder canvasses were back from the repairer and I had the usual problem of remembering which went where. Was the top elevator canvas the narrower one? Was the platform canvas the longest? But at last it was done.

The boss had been threatening to put a drawbar on the binder and hire a tractor, but I pointed out to him that now we had three horses he might just as well keep his money in his pocket, a line of argument always difficult for him to resist.

With everything ready the weather broke, and with each wet day that came and went the boss became increasingly edgy.

We were running out of wet weather jobs. We had whitewashed the insides of every building we could think of, oiled and greased the implements, tidied here and there, and Aunt Kit was starting to drop hints about a coat of distemper in the kitchen.

The boss was given to comparing our sorry lot with the lucky people who lived down south, by which he meant south of the Tweed.

Down south, he would say with great bitterness, they can start harvest and work right through, none of this damned rain. No flattened corn either, and they save a fortune on not having to pay any overtime. Not only did they save on overtime; apparently they didn't have horses eating their heads off when there was no work to do and they were highly mechanised down south with big fields which were invariably flat.

A period of sullen silence would follow, then he would rail against the farmers in the east of the country, who were nearly as well off as those down south. He might for a change blame 'those bluidy Campsie hills,' or perhaps Ben Lomond, for the wetness of the area.

'I've a good mind to put the whole damned lot on a train and go down south myself,' he would say.

'Why don't you?' I asked on one occasion.

'I'm thinking about it,' he had replied darkly.

According to Aunt Kit he had been contemplating this course of action as long as she had known him.

We had nearly got around to distempering the kitchen when the sun broke through. By mid-morning we were able to start opening up a small field with the scythes, hanging the hand-tied sheaves on the hedge to dry. By tea time I had the binder in the field and was ready to make a start.

Donald, muzzled to stop him feeding on the new grain, was next the corn, Blossom on the other side of the pole, and Fanny, fresh and frisky after a summer grazing the patches of short nutritious grass on the moor, was on the outside.

Perched on my metal seat, cushioned with a bag of hay, the horses pulling together, the knife whirring, flights creaking, sheaves spilling out the side, I couldn't have been more content.

Gone was the boredom of the long days scaling dung and wet days whitewashing every wall in sight. No more was I concerned that the boss was grumpy, that people in other parts of the country had all the good weather, big tractors and acres and acres of flat land. The sun was shining, the crop was not overheavy and was still standing, the horses were making light work of it, the knotter was not missing any knots and I was rapidly beginning to forget the weeks of wet weather. I was carefree and happy, and my happiness was not diminished when, after four rounds, the boss came out to rush around stooking as if his life depended on it.

The tea and buttered scones which Aunt Kit brought out were a welcome interlude, and I managed to get back in my seat before the boss could spare the time to have his. I was enough of a tactician not to allow him the chance to suggest we change jobs.

So my idyll continued, watching out for large stones which might have emerged, keeping an eye on the various working parts of the binder, alert at all times to raise or lower the flights or the cutter bar, turning the team at the corners, it was all too good to be true. There was time to observe the rustle of rabbits as they moved deeper into the corn, and the boss trying to do the work of two men or angrily picking thistles from his hands.

With another four rounds done and the boss gradually falling behind with the stooking, I noticed a mass of black cloud astride the Campsies. Soon after I felt spots of rain.

I had stopped near the gate and sheeted up when the boss reached me. 'Bloody Campsies...'

I interrupted hurriedly. 'I'll just take the horses home and come back and help you stook up.'

'Right, and don't be long about it,' said the boss grumpily.

Back in the sanctuary of the stable, the horses wet from rain and their own sweat, I took time out to wisp them down and then sat on the corn kist to enjoy a smoke before I went in search of oilskins.

I had a feeling that, in more ways than one, this was going to be a long, hard harvest.

16
⤝ *Rehabilitation* ⤜

ONCE the rain had started, it attended diligently throughout the harvest. It was what the boss called good spoiling weather. Everything would start to dry up, then down would come the rain.

The binder job had not proved to be the sinecure I had anticipated. Heavy downpours broke the ripe straw. I continually had to back up the horses to clear collections of damp straw and grass off the fingers or off the divider.

Backing the binder on a pole was no easy job for the horses, and I began to wish I had gone along with the tractor idea. The buncher made heavy weather of the broken straw with heads and tails all mixed up. The untidy bundles were difficult to eject, and often came back over to create an unholy tangle. The knotter functioned poorly, twine snapped and spars broke on the canvasses.

I began to wish I was down south where they had big fields, large tractors, standing corn and perfect weather.

The boss's mood swung between suicidal and belligerent. My own patience was a bit fragile and we had the occasional ruck, but, always mindful of the inequalities of our physical attributes, I did most of my grumbling to Aunt Kit or anyone else who would listen.

Soakings were the order of the day as we restooked, teasing apart the growing-out heads to facilitate drying. Finally, having repeated this process many times, we got some of the stooks into rickles in the field and eventually were able to cart it all home.

Stacks, narrow and much propped, allowed for further curing of the less than ideal product.

I could understand the despondency that settled on the boss after the many weeks of soul-destroying effort. He had seen a good crop sadly reduced by the vagaries of the weather. The workload had been enormous, and our labours often rendered futile by the ever-returning rain.

There had been the cost of my overtime and some casual labour recruited to make use of any dry days. Potatoes which had been rogued and inspected and should have made good money as stock seed were still in the ground. Now the boss reckoned, what with the lateness of the season and the amount of rain, they would have to be sold as ware.

The financial implications of all this were not lost on me.

For a few weeks the boss was not much in evidence between milkings. I didn't get my orders in the morning and carried on as best I could. I enjoyed the freedom to plan my work, and, not having to carry the financial burden, was able to apply myself single-mindedly to thatching, hedging, marking out the stubble or whatever the weather or my own inclinations dictated.

The boss's period of withdrawal lasted for three weeks; then he took up where he had left off. I had my orders in the morning with the old thoroughness and attention to detail, but he still seemed not quite his old self. He was quiet, even-tempered, reasonable... I didn't like it one little bit.

The ground had dried considerably, and although the rutted stubble fields bore testimony to the poor underfoot conditions at harvest, the potatoes came out large and sound and bountiful.

'The best crop of ware I've had for years,' opined the boss.

'But you haven't grown maincrop for years, have you?' I wanted to know.

'Oh, I used to,' he said. 'Anyway, if these make a good price when we open the pit later on, they might make as much as if they'd gone as stock seed.'

I could see glimmers of the old optimism showing through; it had to be encouraged.

'And of course you'll have Bertha's pigs to sell. They'll fetch a pretty penny.'

'True, true,' the boss was rubbing his chin, another good sign.

'And you'll have a few stirks to sell,' I went on. 'And with this mild weather and all the rain we had, there'll be grass to fill the cow's bellies for a time, and leave a bit for the outwintered stock to pick at.'

The boss was nodding agreement, his hand rasping busily on his chin.

I dug deep. 'And Aunt Kit's selling eggs at the door.'

'I just wish the beggars wouldn't shite on my car.' There was a querulous note in his voice which I found encouraging.

'At least they make a straight line,' I comforted him with a mental picture of a line of sleepy hens, crouching, droop-lidded, on the roof joist, taking turns to precision-bomb the target below.

'Well, I have to park in exactly the same place or I get another line; and sometimes they manage to plaster the door handle. I think that's deliberate.'

I thought there was the suspicion of a twinkle in his eye, and he was whistling as he walked away. I felt that soon he would be back to normal: belligerent, argumentative, ever-optimistic, mean with his money, pontificating on this and that. It was all very encouraging. However, it was only when I saw him furtively slipping his black hat into the back of the car one day when Aunt Kit was out feeding the hens that I knew his rehabilitation was complete.

17
≈ *The Way It Is* ≈

RAPIDLY shortening days meant the boss was again busy with evening meetings. One I knew to be legitimate was the Young Farmers' Club meeting, which was held in the village school. Whether he attended alone or not was outside my ken.

I had hoped he might suggest that I go with him, but the thought didn't seem to occur to him, although discussion on the speakers had been copious and my interest obvious.

I knew that the club functioned as a two-tier entity, run by Young Farmers who were shadowed by older members in an advisory capacity. The attendance I understood was roughly equal between young and more mature farmers.

'We've got a young chap coming up from Auchincruive to talk about block rotations,' the boss informed me one evening when we were having supper.

'Seems like a good idea,' I said knowing full well what he meant—working things so that corn fields were together in one block and grass and roots likewise.

'Mmm, not always as easy as that. You have to take account of the land.' The boss as usual was looking for the snags.

'I'd like to come,' I blurted out.

'There'll be none of your mates there,' the boss warned.

'I know,' I said. 'But I'd really like to hear it from him rather than getting it secondhand.'

Aunt Kit intervened. 'Take the boy; do him good,' she said.

'Well... we'll see.' The boss seemed reluctant.

On the night, he said, 'I have to go early... a meeting... I thought you might finish off for me and come on when you've finished.'

I had a feeling he wanted me to back out and I was half inclined to use the excuse, but something perverse in me made me say, 'I'll be there, don't you worry.'

'Oh, good,' he said halfheartedly. In the event he had left me more than enough to do.

When I arrived at the school the meeting had started. I stood outside the door and heard the speaker say, 'Ladies and Gentlemen'.

My heart was beating so fast and so loud I felt sure they must hear it inside. Inner conflict raged. The perverse against the cowardly.

I turned the handle and, cheeks hot with embarrassment, stepped inside the room.

The speaker stopped abruptly halfway through his opening joke. The young chairman stared. His older shadow, Mr Jessiman, drew deeply on his pipe, smiled and pointed to an empty seat.

I stumbled to the back of the room and claimed the sanctuary of a shared desk.

The speaker speared me with piercing blue eyes and said, 'I'll start again for the benefit of this young man.'

Just for a moment eyes and broad grins turned my way; then the speaker claimed their interest.

I felt a light touch on my arm. 'Nice to see you.' It was the merest whisper.

I turned and looked into friendly grey eyes. 'Penelope!' My whisper came out as a loud croak.

Heads swivelled and somebody said sssh...

I couldn't trust my voice again, but feeling that Penny's greeting merited a reply, I pulled my notebook and pencil from my pocket and wrote a message straight from the films: 'You look like a million dollars.'

In moments came the reply. 'That's about two hundred and fifty thousand pounds; are you sure?'

Caught unawares, just as I was emerging from my traumas, I was overcome with a need to giggle uncontrollably, to laugh out loud, fall about and hold my sides. As it was I nearly choked trying to control it and ended up with a fit of coughing. Penelope patted my back; Mr Jessiman, enveloped in a cloud of smoke, took his pipe from his mouth and smiled indulgently. A few people looked round, but the speaker, directing a searing glance in my direction, maintained his flow.

Soon I was absorbed by the lecture and took many notes. The boss, seated in front, was also writing furiously. I had a rear view of a well-built lady beside him. They seemed to be on affectionate terms.

Despite the effect on my pulse I took part in the discussion which followed. I could tell the boss agreed with my observation that the big farmers down south would have more to gain from farming in blocks, by the vigorous up and down wag of his head. Mr Jessiman, who at least knew who I was, must have been grunting agreement also, because his pipe suddenly gave off a series of smoke signals.

I had a short conversation with Penny at the end, before she went to join her brother. Some of the farmers' sons greeted me, but politely, without the rumbustious joshing I might have expected at a threshing day or on the football field.

My initiation was over; I found the treasurer and paid my subscription.

At subsequent meetings I found that people kept the same seats, as much creatures of habit as cows in a byre. Penny and I sat together and became friends, swapping books, talking of this and that, at ease with each other, sharing our youth and our sense of fun.

At the third meeting the boss's lady friend accosted me. 'I want to see Fanny's foal when it's born,' she said. I had taken a liking to Peg, as she was called, from the start. A few years younger than the boss, she was substantial and jolly, with a forthright way of coming straight to the point.

'See what you can do with old fat arse,' she said.

I said I would. I could only assume she was referring to the boss and not Aunt Kit.

Life was good. It was great to have a friend of the opposite sex; we had so much in common, were so much in tune. I looked forward to the monthly meetings enormously.

Then one night she was missing.

It was at breakfast the next morning that I came face to face with a fact of life.

'Penny wasn't at the meeting,' I observed.

'Getting ready for her wedding, I suppose,' said the boss carelessly. 'Marrying some chap from away. Ayrshire, I believe. Early spuds, ryegrass seed and a few cows. That's the way it is; farmer's daughters marry farmers.'

He went back to his porridge, unaware of his contribution to my growing up. It was a fact of life in our society.

'Yes, that's the way it is,' I agreed, and with a sigh for what might have been, I set about my porridge.

18
A Piece of Cake

WE were all set for a day's threshing. It was a cold, crisp morning with no rain about and little wind.

'The weather's right, anyway,' said the boss. Over breakfast he was discussing how best to deploy the influx of neighbours who would attend on a reciprocal arrangement.

'There'll be our neighbour from next door; he'll bring Harry with him. Then there's those two rascals you call your friends, from across the way, and the old chap with only half a stomach....'

'Him,' snorted Aunt Kit. 'If it's only half a stomach, it serves him well enough.'

'He can clear up the chaff,' went on the boss. 'You will build the straw stack; Harry can serve you on the stack till it gets too high for him...' The boss went through his plan, fitting the people to the jobs.

My friends Alisdair and Rob, being young and sturdy, would carry the twelve-stone bags of corn off the back of the thresher to the barn.

'How many for dinner?' Aunt Kit wanted to know.

'About fourteen, fifteen at the outside. It'll be tattie soup and tatties and mince, I expect.' The boss was not inclined to be lavish when catering for numbers.

'Of course, just as usual.' Aunt Kit, a veteran of many threshes, knew the drill. Plenty of food, nourishing but not expensive, and making maximum use of potatoes and turnips with some curly greens from the garden.

I was looking forward to my two friends' first visit to what I always referred to as 'my farm'. They had recently arrived to work on a bigger farm nearby, where they shared an outside bothy, and although they had none of the benefits of my 'as family' regime, they tended to have a less restricted lifestyle and were given to playing practical jokes. I had in mind a prank of my own. It would involve Half-a-Stomach and would, if this gentleman performed true to form, take place at the midday dinner table.

Harry was among the first of the helpers to arrive. 'How's the wonder horse these days?' I asked, referring to his pride and joy, the grey Shire. I was on a leg-pulling basis with Harry. In fact, we had been friends since the hay leading. I had warmed to him when he had come over one evening to see how the stacks were 'sitting down', as he put it, and had said that they were better than the ones he had built the year before.

This, of course, guaranteed my friendship, and I took to chatting with Harry's group of older craftsmen in the pub and, as Harry's friend, was accepted as a member by this elite group.

'The horse is fine,' said Harry. 'But they're talking about a bluidy tractor. I don't know where that leaves the horses.'

'Well, the tractors are getting about. You'll just have to learn to drive; anyway, you have an old Standard Fordson over there.'

'We hardly ever use it, but these little buggers with their three-point linkage, well, they'll be stitching up the tattie ground and doing all the ploughing. It'll be the end of the horses.'

Someone passed, lustily relieving himself of intestinal gases. Quick as a flash Harry called after him, 'Cut yourself a piece of cake.' It was his invariable response to any slight behavioural lapse—the passing of wind, the smallest burp, a splutter, something going down the wrong way. The reaction was always the same, automatic, like the 'bless you' after a sneeze.

After a time I picked up the habit. 'Cut yourself a piece of cake, Harry,' I would say if I caught him out when we met up in company, and his toothless mouth between the downward hooking nose and the upward tilting chin would smile in gummy approval.

I often thought how his profile was the nearest thing to Punch I had seen in real life. He was a small man, no more than five feet tall, and nearing the end of his working life. I knew the arrival of a modern tractor would be a worry to him, especially if it meant the departure of his cherished grey Shire horse.

People were beginning to arrive. The threshing tackle was in place in the stackyard, the machinery was ticking over, belt-driven by a giant Case tractor, a living van was parked in a corner and the thresher man and his lad were busy with oil cans. A touch on the throttle and the noise and the dust would signal the start of the day's work.

The boss had delayed having a day's threshing, as he felt the corn needed to cure in the stack as long as possible. Once the stock of old corn was depleted we kept going for a time with odd bits threshed with our barn mill, but the ancient machinery creaked and groaned and complained, and did a fairly inefficient job of separating the corn from the chaff. The giant Blackstone stationary engine which powered the mill machinery was another problem, needing the boss's weight and brawn to turn the great flywheel and coax its reluctant slow heartbeat into life.

The thresher man climbed aboard his tractor and sat importantly, in complete control of the performance. A signal from the boss that the cast was in place, a movement of the throttle lever gradually increasing the revs to working speed, a great cloud of dust as the mill cleared its throat, the first sheaves cut and fed into the drum, and we were under way.

Harry and I had prepared a stack bottom using the thatch that had been stripped from the stacks; my friends had decorated the back of the thresher with sacks to collect the good corn, the tail corn and the rubbish; the boss and our neighbour were busily forking sheaves on to the thresher; two were up there cutting bands; Half-a-Stomach was standing by the chaff shute, armed with a wooden rake and a hessian sheet prepared to carry the chaff to a safe distance for burning.

He wore cord trousers and a donkey jacket, but the obvious quality of his soft felt hat and highly polished boots set him apart. We had all heard about his son who was Chief of Police in Barbados and his daughter who held some sort of academic post in Glasgow, but he was reticent about his own past. It was generally understood that he had spent years in the Colonial Service. Perhaps he had not achieved distinction worthy of mention. Certainly, as a conversation topic it would have come far behind the operation which had left him with only half a stomach.

The day stayed fine and the work went smoothly. From time to time Harry's sharp ears would detect a misdemeanour which called for his automatic 'Cut yourself a piece of cake, boy'. There was much joking and joshing and itching inside collars, as the volume of flying dust testified to the poor harvest and the weathering of the straw and the corn.

My friends, not overworked on the grain side, had time to come over and reveal plans for a major practical joke on the other outside bothies in the area.

New to the district, my friends had quickly discovered that many of the bothies were occupied by older single men who had a record of long service and slept in long drawers and perhaps socks. They referred to such people as 'typicals'.

It appeared that every morning my friends were reminded that another day had started when their boss, with unfailing routine, would hammer on the window and call, 'Oi! Oi! you boys, six o'clock, Oi! Oi!' From a careful study of this early morning greeting, a plan had been evolved to disrupt the tranquillity of 'typical' bothies over the local area.

My friends had researched the morning calls of two other local farmers. The terminology was, of necessity, learned second-hand, but the voice timbre was studied in conversation. Farmer A was old and quiet spoken. Farmer B was brusque and given to the use of encouraging expletives.

These two voices had been well practised, and it was decided that we would attend a dance in Buchlyvie on the Saturday night and disturb the slumbers of our chosen victims on the way home. Meantime I had a little plan of my own which I didn't discuss with anyone.

Half-a-Stomach was not employed by anyone, but turned up at most of the big mill days in the district as a casual worker. It was understood by all that he was not in need of the money but enjoyed the social aspects of the day. At dinner he would sit halfway down the table to have access to as many listeners as possible.

Today was to be no exception: he was first into the kitchen, greeted Aunt Kit with elaborate old-world courtesy and arranged himself in his usual midway position. I hastily ushered my friends into the seats just beyond him where they were immediately embroiled in details of the operation and the result.

With everyone seated Aunt Kit called us all to order. 'Everyone for soup?' she asked.

There was a chorus of assent, with a qualified yes from Half-a-Stomach. 'Only a small helping for me,' he quavered. 'Just half; I only have half a stomach, you know.'

'I think we all know that,' said Aunt Kit grimly, and passed a small helping down the line.

'That's far too much,' said Half-a-Stomach. 'I'll pass it on.'

Alisdair was the recipient of this small serving and Rob was the victim of the next even smaller helping.

Aunt Kit, losing patience, started serving normal platefuls, and Half-a-Stomach, having advertised his misfortune to all, set about a good helping with gusto. Quick to learn, my friends were ready for a game of pass the parcel when the next course arrived, and understood the jockeying for upstream positions among the initiated.

***

Saturday night found us in conspiratorial mood, meeting in my friends' bothy to finalise plans before going on to the dance. Later we were in high spirits, as two hours spent in the convivial confines of the Rob Roy had ensured that the dance had been a great success.

We were inclined to laughter at the slightest excuse as we steered our motorbikes towards our first bothy. Farmer A's voice was projected passably, and there was much cursing and groaning from within. A gap down the side of the blind gave us a splendid view. Inside a match was struck, a stable lantern flickered fitfully and two 'typical' bothy people emerged from their covers. 'Right boss,' said one.

They wore heavy vests and long-johns, and one of them had his socks on. There was much bemused pulling on of trousers, shirts and bib and brace, and it was only when one of them was attaching his pocket watch to his waistcoat with a bootlace that they noticed the time. Another watch was consulted. Yes, it was two in the morning—and it was Sunday. The explosion of bad language from within masked our uncontrolled laughter without, and we withdrew undetected.

With the residues of our alcohol intake still circulating and nearly helpless with laughter, we threaded a perilous path to our next victims. Here a deep voice was needed. A heavy hammering on the window and a 'Come on, you buggers; get your feet on the floor' had these two typicals up and dressed like a shot. It was the same as before, except that when we left, our victims were trudging off to the byre. I felt a little sorry for the poor cows.

Our manoeuvres continued for some time, but inevitably the joke began to pall. Anyway, our tendency to boastfulness eventually stirred up a lobby against us, and we finally decided to discontinue our activities when my friends left their beds at two o'clock one Sunday morning in response to an 'Oi! Oi! Come on you boys. Oi! Oi!' It seemed the 'typicals' were fighting back, and a draw would be a good result.

The tractor arrived next door the day Harry went into hospital. There was some talk of horses having to go, but Harry was past caring. The cancer reduced him rapidly, and when I saw him in hospital I knew his time was short.

As I was leaving, a sob rising unexpectedly caused me to gulp.

I heard his voice, faint but resolute. 'Cut yourself a piece of cake, boy.'

A few days later I rushed to the hospital in response to an urgent call. I was too late. He was being tidied up.

I went into his cubicle. There was a frill round his neck. His mouth was a round toothless 'O' in the small gap between his nose and his chin. Two attendants had nearly finished. As they moved him slightly, air passed from his body.

My response was as instant and natural as if we had been enjoying a pint in the pub or building a straw stack together.

'Cut yourself a piece of cake, Harry,' I said.

An old (and rather battered) picture of the author with Donald and Blossom in competition trim

The boss with two young friends, on a horse called Nancy, replaced some time before the author arrived

The boss and helpers take a break in the hayfield

The author as a young man

The author in his working clothes, sowing fertilizer the old way

The boss horse-hoeing roots before the time of the book

The ubiquitous Standard Fordson, a good 'workhorse' but over-taken by more sophisticated tractors with their own hydraulically operated implements. As these proliferated, horse numbers declined.

The boss today, at home in Drymen

The author with his wife Renée, still in double harness after 54 years

19
⁓ *Winter* ⁓

*T*HE winter brought snow with all its attendant problems. With the countryside an Arctic wasteland and minor roads impassable, we had to transport the milk cans, wobbling precariously on improvised sledges drawn by Donald and Blossom, to meet the milk lorry in Balfron.

This to me was a great adventure, but the boss didn't seem to share my enthusiasm, muttering about wasted man-hours and inefficient use of resources. A bowl of soup in Balfron's restaurant, with the horses nosebagged and tied to a lampost, was for me pure Wild West.

It hadn't turned out to be the best winter for minding sheep either, and the boss had taken in hand a hundred Blackface down from a hill farm.

Even before the snow, I had my work cut out freeing them from briars and the like. Now, with the low ground deep with snow, we had moved them to the high moor where the wind had bared considerable areas. With a bit of hand-feeding each day they were surviving well enough.

Bertha stayed snug in her sty. She was showing that nice curve to the underline that denotes mid-pregnancy. She had failed to hold to the first service after weaning, but luckily for her, the boss was too preoccupied at the time to notice. Aunt Kit had arranged for another visit from the boar, and the success of this union saved Bertha's valuable carcase yet again.

Snow was still on the ground when I had word from big Jake that he was ready to start breaking in his young horse. If I could attend on the next Saturday afternoon, I could help him. The young mare, rising three, was a dark bay, a really good type, with a nice lively action and a proud carriage of the head.

Jake explained that she was well used to standing with her harness on and had been driven in long reins, so we had reached the point where she would be required to pull something.

Jake had organised a section of tree trunk which would sledge along in the snow and not be too much of a pull on her shoulders. A chain fixed to the rear was to be held by Andy, who would act as brake man in case the log should jerk forward and catch the mare's heels. Not that this was likely, as it was hitched well back on a long chain. My job was to be on a lead rein on the off side.

We stood around in the stable, talking and smoking.

'Her name's Bess,' said Jake, going up to fondle her ears. 'Come up and talk to her; get her used to your voice.'

I made a fuss of her, running my hands down her legs and lifting her feet, touching her head and talking.

Bess showed no sign of nervousness.

'She's been well handled, Jake,' I observed.

Jake smiled. 'Since she was born, near enough. I think you've guessed the secret—what makes this part of the job easy.'

'Handling the foal?' I suggested.

'Aye, you've hit the nail on the head. It's hellish important. Get it used to the halter, take it for walks, pick its feet up. When it's weaned, be its mother; it'll be glad of the company when it's left on its own.'

Bess stood quietly while Jake hooked her chains to the swingle-tree. There was a stout lead rope to each side of the bridle.

'Take the offside rope, keep well back and only pull on it if she breenges to my side, or tries to bolt. I'll try leading her close up for a start. If she tries to go, I'll stand back and we'll both hold back on our ropes.'

Jake had chosen an open route which allowed for manoeuvring space in case of trouble. It was a track leading to a field, unfenced, with snow-covered grassland either side.

We set off, Bess prancing with short, springy steps as she felt the load pull against her collar. Jake held firm, talking to her, his voice calm and soothing, reaching up to take hold of an ear as Bess bucked a little.

The path led gently upwards. After half an hour Bess settled to a steady walking pace; we struck off across a grass field to give her a more level pull and then turned downhill to give her a feel of that.

'Just watch that log behaves itself, Andy. If it touches her heels there'll be hell to pay. And,' he added darkly, 'you'll be in the middle of it.'

Ten more minutes and we stopped for a breather. Jake pulled a handful of oats from his pocket and rewarded Bess. 'Fine so far', was his verdict.

When we set off again I was driving Bess on long reins and Jake was still near her head. By the time we'd been out two hours, I still guided her on the long reins and Jake had left her head, but I was surprised at how tired she had become. She was beginning to roll a bit on her collar.

'She's knackered,' said Jake. 'She's ready for a rub down, some corn, and a good rest.'

'But she's done nothing, not compared to what working horses have to do.' I was surprised at what appeared to be a total lack of stamina.

Jake laughed. 'Think back to when you were fourteen. How

much work could you handle then? The mare's got a lot of growing to do yet. She's immature still and needs to get used to working.'

'So what's next?'

'I'll give her some work tied back to an older horse, but not too much, no more than two hours a day.'

'And then the cart?' I suggested.

'That's it, in a few weeks. I'll be sending word.'

This proved to be an equally quiet affair. Jake's boss came into the stable as we were getting ready. He was carrying a shaft of wood with a string loop at the end. 'You might need the twitch, Jake,' he said.

'Aye, right boss,' said Jake. 'I don't think I'm going to need it, though.' He slipped the string loop over his wrist and carried on with his preparations.

Bess had been working in chains most days, so was hardly fresh. She didn't quiver as we dropped the box cart on her. It was much as before with Jake and me either side on lead ropes and Andy riding in the cart to drive Bess with long reins. The initial rush, the rear and the neigh were token. Bess soon settled down to the plodding walk that was to be the hallmark of her working life. We were all three riding in the cart when we returned to the farm.

Back in the stable I saw the twitch which Jake had tossed in a corner. It had never left the stable.

I felt I had learned a great deal about gentling a horse.

'I'll buy you a pint, next time I see you,' I said.

'Aye well, it might cost you more than one,' Jake grinned. 'I'd best be getting on.'

I left him then. He had a horse to rub down, and reward with corn, and make a fuss of.

Despite the weather, the boss and I achieved perfect attendance at the Young Farmers' Club, although attendance at some of the meetings was sorely depleted and local speakers had sometimes to fill in for others from 'away'.

The boss showed great ingenuity in respect of transport. At the worst of the weather we donned our wellies and walked in a straight line to the village. Hedges and stone walls had disappeared under drifts, so our passage was unimpeded.

'The boots will get a good clean by the time we get there.' The boss could find a plus point when it suited him. When travelling by road became less hazardous, he borrowed the tractor from next door.

With equal determination his lady friend, Peg, seemed to be ever-present and we were rapidly becoming friends. I enjoyed her irreverent sense of humour. She was nothing if not direct.

More than once over the winter period she mentioned her wish to see Fanny's foal. 'I'm sure you will be invited to tea,' I reassured her. 'I'll see what I can do.'

I felt I was moving across social frontiers. I was picked for quiz teams, debates, and at the end of the season I played right back for the football team, managing to score an own goal. Since we were playing the redoubtable Rovers and were at the time five-nil down, it hardly mattered.

At the meetings I was intrigued by the presence of a tall, distinguished-looking middle-aged man who always sat at the front. His speech, his clothes, even the way he held his cigarette set him apart. He was indeed a 'Sir', with a large house, a home farm and a manager.

One evening he spoke to me, asking me about my future plans, and what I was doing about them. Overcome with shyness in his presence, I felt I acquitted myself rather badly.

'You ought to read,' he advised. 'Learn as much as you can, then try to get to agricultural college. There are scholarships and bursaries, help of one sort or another.' He moved away to talk to the boss.

Soon after, I was presented with Watson and More's tenth edition of *Agriculture*. Since it was past the November term and some way off the May term, I could not regard this as tactics on the boss's part. I was enormously pleased and read voraciously.

After a time I began to query things with the boss: should he perhaps be steaming up a bit more, had he considered this, had he thought of that... I was full of ideas.

The boss intimated that he was delighted that at last I had noticed the dairy herd, but if I got too big for my britches he would take the bluidy book off me.

Aware that the boss didn't indulge in idle threats, I kept my own counsel, and my book, hoarding the treasure-house of knowledge I was discovering between its pages against an unknown future. I had long admired the boss's business acumen, his instinctive feel for the optimum between input cost and expected return. He had been doing the job for a long time now. The whole farm business, the husbandry that ran a parallel course, all that was continually being thought about, balanced one against the other with simple logic, while keeping abreast of future trends.

I knew he had been thinking about a tractor for some time now, trying to justify the expense and not unaware that the winds of change were perhaps reaching out to touch his small farm.

I was in the stable one day when I realised that a certain restlessness I had been suffering from was ambition. Perhaps it was time to move on to greater things.

I hopped up on to the corn kist to smoke a cigarette and think this one through.

The horses moved restlessly, making little sounds, chomping at their hay, looking round to see what I was getting up to. Fanny in the middle stall was getting big.

There was her foal to look forward to in the spring. Donald was in fine fettle as usual. Blossom needed my care and attention.

I had a lot to learn from the boss and he might need my opinion on the tractor issue; then there was Aunt Kit's cooking, and I would miss the horses.

I would move on, but not yet. I hopped down and set about my chores.

20
⤳ *The Visit* ⤳

*F*ANNY foaled outside in the paddock without any trouble. The foal was already up on wavering legs and finding the teats when the boss and I went out first thing to check.

'By far the best way,' said the boss. 'It's how nature intended it to be; the mare can exercise up to the last minute. Less chance of infection too. If she'd been in a loosebox you would have been checking every five minutes; then you'd have started to worry, wanting the vet, I shouldn't wonder. As it is she's foaled in her own time, you've had your sleep and I've maybe saved a vet's bill.'

I had to accept that his rationale was accurate as far as my worries were concerned. I had argued for the loosebox, disinfected and clean-strawed, with frequent inspections and the vet a phone call away.

This was my first foaling; I had wanted to be there at the birth, but looking at the fine foal Fanny had produced without fuss, I was just happy with the result.

'Your Peg wants to see the foal,' I reminded him.

'Oh, ah, so she does, that's right.' The boss seemed less than keen on the idea.

'I told her you'd most likely ask her to tea,' I went on remorselessly.

'Oh, you did,' he said numbly. 'I would have to make some sort of arrangement with Aunt Kit.' He stood rubbing his chin, pondering his problem; then he spoke suddenly. 'Bugger it, I'll talk to her now.'

He stomped off. I watched him head for the kitchen door, slow as he neared it, then stand rubbing his chin for a time before making for the byre. Either his courage had failed him or he had replanned his strategy.

At supper that night he was more than usually talkative, going over the pros and cons of the past year.

The bad harvest had an innings, but this was balanced off to some extent by the bumper crop of late potatoes, a commodity he hadn't produced for some years. Somehow he managed to give the impression that this year's crop was the result of foresight and forward planning, as if a poor harvest had been taken into account.

The early spuds also were profitable, with the merchant shouldering most of the risk. A mild autumn had kept the grass growing. A few agisted sheep over the winter had been grist to the mill. Fertility had been good in the herd and the autumn calvers had calved on time, catching the good milk prices.

This also was a triumph of planning over the hit-and-miss methods of some others—I think he was referring to our neighbour—and Bertha's pigs, sold (except for the one which hung above us) at seven months of age, having achieved great size, had made a welcome contribution.

I felt that he was working up to something, although since my elevation to membership of the YFC and the friendship I had struck up with his lady friend Peg, he often discussed business with me.

'And now we have a fine foal, yes, a very fine foal. It was a bit of luck finding that mare, didn't cost a lot either....'

I waited. Now he would ask about Peg coming to tea.

Aunt Kit maintained an unhelpful silence.

'Ach well,' said the boss into his plate.

'The lady who sold us Fanny wants to see the foal,' I said impulsively. 'I see her at the YFC.'

'Oh, you mean that Peg.' Aunt Kit spoke through tightened lips.

'Er, well, I thought we could offer her tea,' said the boss, finally grasping the nettle.

'It would be only civil,' said Aunt Kit formally.

'On Sunday,' I suggested feeling the need to lead the conversation.

'That's it, that's fine,' said the boss, 'If it suits you, Aunt Kit.'

'I'll make it suit,' said Aunt Kit ungraciously.

'Sunday then,' said the boss, and the relief in his voice was evident.

I was in the stable when Peg arrived at the wheel of a rather elderly long-nosed Ford Ten. Through the window I saw the boss come out to greet her.

Peg was smartly dressed in tweeds and brogues. She was smiling broadly as she greeted the boss, before slipping her arm through his, as they went to confront Aunt Kit.

I wondered what she called him in their private moments. When she mentioned him to me in her boisterous, good-natured way, it was usually 'old fat arse'.

Later I escorted the party, including Aunt Kit, to the paddock. Fanny approached, her foal, stilting on long legs, close to her side. It was beautifully marked, with a generous white blaze, four white feet and a coat of dark grey.

I conducted Peg on a tour to see the foal from all angles.

'A nice colt, it'll be black like its dad,' she reassured me. The little mare looks well too. I should have charged old... er, your boss, a bit more. He caught me off my guard.'

I had a momentary vision of the boss clinching the deal in the car, parked in some remote gateway, but put the thought away as unworthy.

There were a few jobs needing my attention; the others went in to tea in the parlour. The best china would be out and Aunt Kit had been baking.

I next saw the boss when he came out to finish a few jobs, leaving the ladies washing up and chatting in the back kitchen.

'How are they getting on?' I asked.

'Oh, fine, fine.' The boss seemed a bit sheepish, ashamed perhaps of the moral cowardice that had delayed this meeting for so long.

Later that evening, when Aunt Kit and I were alone, I asked how she had got on with Peg.

Aunt Kit positively beamed. 'Oh, I did like her. She's so jolly, and you know, she doesn't intend to get married for a few years. She feels she's needed at home and is quite happy with things as they are. I shall probably be able to draw my pension before then.'

Aunt Kit paused, then went on. 'There's just one thing.'

'Yes?' I prompted.

'Well, she has a very funny way of putting things—what she calls the boss—you'd never guess.'

I had to laugh. 'I have a pretty fair idea,' I said.

21
⁓ *The Tractor* ⁓

*T*HE tractor debate carried on through the spring, and finally, just as I was giving the turnip ground a final drag with the spring tine harrows, a Ford Ferguson arrived. It wasn't new but good secondhand, and it came with a two-furrow plough, a ridger and a cultivator.

It hadn't been an easy decision for the boss to make. He had done a lot of thinking out loud, and I, having read the farming papers and generally given the subject some study, was a useful foil for this activity.

We had been sitting on after supper one evening, while Aunt Kit cleared away, and the boss was marshalling his thoughts.

'In the first place I have never borrowed from the bank except for seeds and fertiliser, which gets paid off by the harvest. In the second place I don't like paying a big sum out of our working money in case I should need it... for your wages and such like.' The boss had a way of personalising his arguments.

'You don't like parting with money, big or small,' interjected Aunt Kit.

The boss let that one ride, and pressed on. 'The trouble is, I can't see any extra coming in just because we lay out money on a tractor and a few implements. We should really sell two of the horses... Not that I would consider it,' he added hastily.

I thought it was time I contributed, so I said, 'Down south, farms are getting mechanised up to the hilt. Even our neighbour is talking about silage, which means buying a buckrake. It makes sense when you think of the weather we get sometimes.'

'That's just it. Once we start gearing up to the tractor it'll go on and on: a buckrake, a trailer, a reaper. Still, I suppose we could convert some of the tackle, the corn seeder, the fertiliser distributer, maybe the reaper; anyway, it's bound to come sooner or later.'

And come it had. Perhaps the decision had not been entirely based on pure economics. I thought perhaps my reference to the neighbour's forward-thinking attitude might have had a bearing.

A cart gave way to the smart grey tractor in the cart shed. We circled it admiringly, sat on it, started it up and listened to the sweet sound of the engine; but for three days we found no particular use for it.

It was during the next day that the boss fell ill. I knew it was serious when he let Aunt Kit persuade him to his bed. My worst fears were confirmed when the doctor arrived in the afternoon.

Remembering events next door when Harry was taken to hospital just before their tractor arrived, I was gripped by a sick fear. Could this be a repeat of those sad days? But the boss was in his prime, as strong as a horse; I hung around and caught the doctor by his car.

'He's strained a valve of his heart is my guess,' grunted the doctor, an elderly, taciturn man with a direct approach and little in the way of bedside manner. 'Silly young bugger, just because he's as strong as a horse doesn't mean he has to work like one. Next time you see him lift a two-hundredweight sack on his own... but that'll be a while yet; he'll be flat on his back for weeks, maybe months. Oh, don't cause him any worry about the farm, take it on, son.' He patted me kindly on the shoulder. 'Back tomorrow,' and he climbed stiffly into his car and drove carefully away.

The boss was in a bedroom downstairs. Aunt Kit ushered me in. At the time it didn't register that this was the very first time I had been through the door which separated the kitchen from the rest of the house.

He seemed smaller somehow, sunk deep in a feather mattress, sentenced to weeks of immobility, vulnerable, as helpless as a baby.

'You might have to help Aunt Kit turn me at times. I don't want to get bedsores on my arse.' His attempt at a smile twisted a knife in my heart.

'I won't get my arse kicked for a few weeks by the look of things, boss,' I said.

'You might have to kick mine if I start feeling sorry for myself.' He smiled weakly. 'The farm's yours for a bit, lad. You'll have to do the milking; that'll be a new experience, and you might have to get that new tractor dirty.'

The next morning I dragged myself from my bed an hour before time. Aunt Kit was already in the dairy assembling the machines. I got on with feeding concentrates and washing a few udders.

Despite only limited experience in the byre, I knew the routine and could manage the milking well enough, and I had the milk cans on the stand ready for collection in good time.

With the horses out to grass I had no stable duties, so when I had finished cleaning down the byre, I was ready for breakfast.

My next job needed no planning. The turnip ground was fit; it was dry overhead, and the tractor was equipped with a three-row ridger. I checked that there was petrol in the small tank for starting, paraffin in the big tank and oil in the sump. I started up and managed to attach the ridger on the three-point linkage.

It was at this point that I was assailed by a crisis of confidence. Could I get a straight start? Would I control the depth?

Donald and Blossom were not far away. I opened the gate and called them in for a bite of corn.

In the field I set up my poles, and keeping a line between the horses' heads, ridged out a shallow marker. With the horses either side of it on the return trip, I deepened the furrow, taking out any wayward kinks in the process. A few more ridges and I was happy with the result, but had had time to ponder on the amount of work I had to get through.

I left the ridging plough under the hedge, took the horses back to their field and set about the job with the tractor.

I soon found that with one set of wheels and one ridging body running in my last furrow, I maintained perfect depth and could hardly deviate from the straight.

With two ridges forming at a time, and with the throttle set to far exceed horse speed, I began to see where my salvation lay. I was achieving much more output with far less effort, and I certainly didn't miss trudging, feet in alignment like a Red Indian brave, up the narrow vee of the ridge bottoms.

That night I was able to report that the field was all stitched. I think the boss was impressed.

Peg was a regular caller. She took over the book-work and generally dealt with the business side of things. I was impressed by the patient, gentle way she handled the boss, cradling him while he took nourishment, reading the farming papers to him, helping Aunt Kit with bed baths, shaving his chin while brightening his day with local gossip.

'Always was good with sick beasts,' she said to me, but her smile was forced, and her eyes were anxious. 'He'll be up to his old tricks soon,' I reassured her, but looking at him, helpless in the big bed, I wouldn't have taken bets on it.

My evening visit to report on the work seemed to brighten him up. I was reassuring about the land work, and threw in a few mishaps on the byre work to make him feel better. Indeed I wasn't too keen on leading out the bull to do his work. When I had seen the boss lead him on a piece of rope, they had somehow seemed well matched—two amiable giants out for a stroll. At close quarters the bull seemed huge and full of menace. I always used a pole and hoped the distance between us would prevent my fear from being communicated.

With the help of the tractor I kept abreast of the field work. Pulling down the potato ridges and stitching back up three at a time was a revelation. Later, cultivating three rows at a time between the plants made horse-hoeing seem archaic.

I was becoming more and more interested in the dairy herd and felt we could push the nine hundred and ten gallon average up a bit. I never quite established rapport with the bull and would have been happier with a younger and less massive animal.

It was a month before the boss was about in the house, and we had the hay up in pikes before he appeared outside. I had found two casuals to help out, making the pikes, with Donald and me sweeping in with the 'tumbling tam' and Aunt Kit and Blossom raking after.

The boss was in the yard looking thin and frail one afternoon when I rode in on the tractor. 'Hold on a minute,' he called, and stepped up to stand on the drawbar. 'Nice and slow,' he said. 'The grand tour.'

We covered the farm field by field, and as we progressed I felt the boss gradually come back into the land of the living. His interest and enthusiasm glimmered, if fitfully, and I now felt sure that time would heal his physical weakness.

We finished with the dairy herd where my new-found interest must have shown.

'I don't suppose I'll get my dairyman's job back now,' he said, with a semblance of his old banter.

'Well,' I responded. 'If we went AI I might consider carrying on. As it is, well, that great bull is missing you.'

'Ah,' said the boss. 'Now that might just be an idea. AI might be a way of improving the herd...' and he was away, thinking out loud, planning for the future, how to improve things, how to move forward, and I was listening, and learning, and putting in my halfpennyworth. It was like old times and I couldn't have been happier.

Over the next two weeks the boss gained strength quite remarkably, and on the Sunday, Peg came to tea.

I was sitting in the kitchen when she came out to sit with me. She put a friendly arm round my shoulders and said, 'Old fat arse won't know how to say it, but he's ever so pleased with the way you've handled things. He wants you to join us for tea.'

I was pleased on two counts. Peg's boisterous use of the descriptive endearment, unheard during the illness, was a signal that the boss was better; and I was about to enter the parlour for the first time.

So another citadel was about to be invaded. I would cross another threshold and another social barrier would be down.

22
The Band

I heard the phone ring. The boss answered, then poked his head round the kitchen door. 'It's for you.'

'Who can it be?' It was almost unheard of for me to have a phone call, but of course, now that I had extended my terrain beyond the boundaries of the kitchen, the phone was no longer exactly out of my reach.

'Oh, one of those smart-arse friends of yours from the Young Farmers' Club. He probably wants somebody to make up the numbers in a stock-judging team or something.'

'I don't think so...' I began, but the boss interrupted. 'Go through and answer the damned thing and you'll find out.'

The boss, now fully recovered from his illness, was his usual unpredictable, often irascible, self. Being well outside the age limit for YFC competitions, he had perforce to count himself along with the older associate members, and had a poor opinion of some of the sparky younger fraternity.

I 'went through'. It was a farmer's son with whom I had struck up a friendship during last winter's activities. We had turned out to be the stalwarts of the debating team.

Would I like to go with himself and another chap to Greens Playhouse Ballroom in Glasgow on Saturday night? There were apparently acres of dance floor and hundreds of unattached girls. We wouldn't have any trouble finding partners... and Joe Loss was playing.

I didn't hesitate. 'You're on,' I said, and we quickly made arrangements.

I had never been inside a large city dance hall and had never seen a famous band in the flesh, and Joe Loss would probably be waving his stick at an eighteen-piece outfit.

When I explained the burden of the phone call to the boss he surprised me with his 'big band' knowledge.

He had heard Geraldo at the Glasgow Empire, Oscar Rabin in Kilmarnock; he had worn out a pair of shoes dancing to Jimmy Shand when he came to Aberfoyle... oh, and he had gone to Edinburgh once to see Joe Loss.

'He'll be wearing a red carnation and waving a longer baton than most. And he'll have more Brylcreem on his hair than even you have when you're off out.'

The adventure started with a bus ride into Glasgow. We were full of boisterous good humour, best suited, well scrubbed, with slicked-down hair, and an inner feeling of confidence which evaporated a little as we paid our money and entered the vast hall. The sheer size of the place, the lighting, so many people, and not a soul we knew. It was a far cry from our usual Saturday night thrash.

We stood on the fringe of the action. I could see my friends were as overawed as I was. 'Maybe we should have a drink first, till we get our bearings,' I suggested.

The others nodded mutely and we made our way upstairs to the gallery. After we settled at a small table overlooking the dance floor, my friend of the debating group, Alan, was the first to break the silence. 'It's a hell of a big place,' he managed.

Bob, not a debater and therefore not expected to bring argument to the conversation, agreed. 'Could get our village hall in here two or three times,' he said in tones that told me he would have been happier at his usual Saturday night hop.

There was some activity down below. The resident band which had been playing came to the end of a number, there was a ripple of applause, then an interval was announced, and as people began to drift to the bars, the lights moved off the stage as the musicians removed themselves and their instruments.

When the lights again found the stage, the Joe Loss Orchestra was in place, smiling, instruments at the ready. Someone counted them in and they struck up a few bars, fading as the spotlight moved to the side of the stage to illuminate the entrance of the great man himself.

He strode briskly into the bright pool of light which lit his way to centre stage and raised his baton to bring the orchestra to the ready. On the down beat the sound crashed out, and Joe Loss was turning to smile at the fans.

It was magnificent, thrilling, and much as the boss had described it: the long baton, the red carnation adorning the immaculate evening dress suit, hair polished like patent leather, and a smile ever ready for the dancers passing by. Then the spotlight picked up the vocalist coming forward to perform. She wore a dress that shimmered silver in the light and her voice, amplified to reach every corner of the room, spoke of things I had been too busy to give much thought to recently—love, romance, excitement.

'Better get a few pints in and have a dance or two. We can't sit here like lemons all night.' Bob was obviously mindful of the fact that girls had figured in our original plan.

'The pints seem like a good idea anyway,' said Alan with a nervous glance at the crowded floor below.

We were all feeling a little bit out of our element; we were country boys, unused to all this glitz and glamour, but a few pints helped to restore our confidence.

We descended to the dance floor and surveyed the scene. Our experienced eyes soon told us which were the unattached young ladies. It only remained for us to walk up and secure a partner.

The available females seemed to congregate in groups giggling and talking together. This made an approach seem more difficult, and another dance finished without us having made a move.

When the music struck up for a quickstep I found myself standing in front of a tall girl with blond hair and limpid blue eyes. I had pre-selected her because she seemed less daunting than some of the others in her group.

'Er... would you care to dance?' I asked formally.

'Hear that?' she shrilled to her mates. 'Manners, right laddie, I'm yours,' and she stepped close, into my arms.

In fact she was limpet close and I found it hard to get into my stride. We came to a halt.

'Don't you know your right leg from your left?' she wanted to know, and gave my left thigh a tweak. 'That's your left, sonny.'

As we got under way I said, 'Sorry about that, I usually do better at our village hall dances.'

'Ah,' she said. 'Up from the country, eh!' She made it sound as if I had come up from some deep, mysterious pit, an alien perhaps just arrived in a flying saucer, but whatever, having established that I was not of the Playhouse, not even of Glasgow, she seemed more prepared to make allowances for my deficiencies.

'Steer me over by the stage,' she demanded.

I was performing adequately now and was able to get close, turning her so that she could bask in the smile from above. I saw a limpid blue eye close in a wink but it went unnoticed. The smiles and nods were as professional as the bouncy athleticism of the conducting. It was a great performance.

The dance finished but the limpet was not finished with me. She steered me back to her group where I was trapped until the music started up again. 'Here, Maisie,' said my erstwhile partner. 'This lad's up from the country. He can dance a bit though he sometimes forgets which leg is which.' She tweaked my leg again fairly high on the thigh and to the inside. 'That's the left, you start off with that one.' Maisie claimed me, and we were off to a reasonably smooth start.

'My father was in the army,' confided Maisie. 'A sergeant major, so he was.'

I expressed interest.

'He had trouble with some of the lads up from the country.' I made no comment, so she continued. 'Yes, some of them didn't know their left leg from their right. It made drill a bit difficult, so it did.'

I felt she had paused for a show of interest so I said, 'What did he do about it?'

'He got them to think about one leg as "straw leg" and the other as "hay leg"; then instead of shouting "left, right," he called "hay leg, straw leg". Of course most of the country boys were alright; it was just a few. Anyway, I think you're alright. The girls were just teasing. It gave them a chance to pinch your leg.' She gave my leg a friendly little tweak as the music stopped.

I had been entrapped by the unattached females for an hour and felt the need to escape. They had shared me around with great fairness, pulled my leg, pinched my leg and generally had their money's worth. After a shaky start I had held my own, teasing about town girls—goodness, never had a roll in the hay? You ought to try it sometimes—and I was pinching back with the assurance of a gigolo.

But enough was enough, and my chance came after a wild jitterbugging session with Trixie. Much shaken up from my vigorous handling, she said, 'I think I'll have a wee sit and a smoke,' and made a beeline for the ladies room.

I slipped up the stairs to the balcony where I found a seat overlooking the band. I had a perfect view of the Loss hair-parting and, more importantly for me, a perfect bird's-eye view of the drums.

Recently I had been taking an interest in drum kits and drummers, and our local band drummer had taken to handing me the sticks when most of the musicians sloped off for an interval drink. I was under no illusions about this. Left unattended, the drum might have been at the mercy of a few drunks. I, being the lesser of two evils, got a chance to practise.

For half an hour I watched the scene below, fascinated, the swish of the wire brush in the drummer's left hand rotating on the snare head while the other brush flickered here and there, finding a cymbal, tickling the top hat, reflecting the light; and the drummer himself full of rhythm, body swaying, his head moving to the beat of a slow foxtrot.

Bob found me and reminded me that we ought to be heading for Buchanan Street, as the last bus would soon be leaving.

We went in search of Alan and discovered him in a toilet, fiercely debating the merits and demerits of a certain well-known football team with a pair of rough-looking individuals who seemed to be on the verge of settling things in the traditional manner. One had a grip of Alan's lapels as we entered.

'He's got a bus to catch,' I explained. 'It's been nice meeting you; maybe some other time.'

Mollified, one of the roughs explained. 'He just wouldn't let go, kept on talking. I was just about to shut him up when you walked in.'

'He's a bit of a debater, might make a politician some day,' I explained.

'Maybe I should have given him a bit of a bash in that case. Anyway, get him off to his bus.' Gratefully we took our leave.

On the bus we compared notes. All in all it had been an eventful evening. It was as we parted that I suggested forming a band.

I knew full well that most farmer's sons had been forced to have music lessons as children. Of these two I knew for a fact that Alan played the violin and Bob could perform on the piano accordion.

'But you don't play anything,' said Alan.

'The drums, sometimes, in McFarlane's band.'

'Only in the interval,' reasoned Bob.

'It's a start. I could learn,' I said, and a seed was planted to bear fruit at a later date.

23
⇜ Change ⇝

*B*Y harvest time the boss was fully restored and playing his usual stalwart part in proceedings.

The tractor also was giving sterling service. Draw-bars were appearing on many implements, including the binder. This, of course, created a two-man situation or, in this case, a man and woman situation, because Aunt Kit became part of the team.

Given the choice between sitting on the tractor or climbing to the more precarious perch on the binder, she chose the former. After minimal instruction she quickly became adept, if a little heavy on the throttle.

I was glad the horses were spared this onerous task, and well aware that they were figuring less and less in the daily life of the farm. Change was in the air, and the horses were spending more and more time peacefully browsing the expanse of the moor.

A trailer with harvest frames had been purchased after the usual lengthy debate, and other items were on the agenda. A brick silo had been built during the summer to store distillers' grains, which tended to be either in a glut situation or hard to come by.

This had probably been paid for by Bertha, who had finally paid the penalty for a litter of six followed by a failure to conceive. I heard the boss tell the abattoir people, when he booked her in, that they would need a special high hook to hang her. Fanny's foal had turned out to be black, as Peg had predicted, and stood high for his age. I had ideas about talking the boss into showing at a later date.

But now the harvest was over, with much less bother than last year, and I was backing the trailer into the midden.

For the first time I realised how much I was going to miss the horses. All the pressure of time during the busy seasons had made me grateful for the speed and handiness of the tractor. Now, with the annual moving of the mountain in prospect, and with no deadlines, I looked at the inanimate grey machine with distaste.

I couldn't groom it in the morning, feed it hay and oats, clear away its dung, talk to it during the working day. I missed the smell of the stable, the sight of the horses fidgeting as they munched their hay, the sound of their hooves on the cobbled floor, and the sight of them in tandem hauling their cart up the rise, with power and technique.

The boss was passing with a rattle of milk cans. 'You'll do it in half the time with the tractor,' he called out cheerily.

'I'm missing the damn horses,' I said grumpily.

'Ochone, ochone,' he said. 'And you still living in the past.'

I had to look 'ochone' up in the dictionary.

24
≈ *Rainy Season* ≈

'PERHAPS we shouldn't be out here,' said the boss. 'What will the neighbours think?'

It was November, we were shawing turnips, the rain seemed set in for yet another day and the ground beneath our feet was rapidly taking on the appearance of a paddy field.

'They might think we're planting rice,' I suggested. 'Then again, they might think you've gone funny in the head and I have no choice about it.'

The boss straightened, giving his back a stretch. There was something prehistoric about his appearance. His legs were wrapped thigh-high with potato bags bound round with twine. A large, heavy-weave corn sack covered his shoulders and was held together at the front with a nail. He looked bulkier than ever with an assortment of layers beneath the cape, and the whole sodden and bedraggled, with quantities of soil adhering to the fronts of the thonged legs.

'Well,' he said slowly, giving his chin a good rasp with a wet and muddy hand. 'I'll give you some choices.'

'Go on,' I said. 'There can't be anything worse than this.'

'Tidy the implement shed and get everything in the right order for the jobs.'

'Done it,' I said. 'Three times in the last month.'

'Grease the tractor and implements.' I thought of the tractor and implements oozing grease at every joint. 'Done it.'

'Get the spiders' webs down in the byre.'

'You did that,' I reminded him.

'You could tidy up the stackyard, pick up the props that are lying about, stack them tidy. At least the neighbours wouldn't see you.'

'I would still be getting wet,' I pointed out.

Then the boss slapped his thigh, dislodging a large quantity of mud. I've got it,' he said. 'You can paint the kitchen for Aunt Kit.'

'I thought you were going to do that sometime,' I demurred.

'Well I was, sometime, but you can do it. It's a nice dry job,' said the boss generously. ·

'I might make a mess. You know what Aunt Kit's like...'

'Well, it's up to you. I can't say fairer than that.' The boss was obviously well pleased with his negotiations.

'I think the rain's slackening,' I observed brightly. 'We could stay out for a bit; we're wet now anyway.'

'Just one condition,' said the boss making full use of his advantage. 'Any more questions about those mangelwurzel things and you'll be in working for Aunt Kit.'

'I'd still like to know the answer,' I said. 'It seems funny to me that Aunt Kit uses salt to kill weeds in the cobbles round the door and people feed salt to their mangels as if it were potash.'

'The answer is to grow turnips,' said the boss shortly.

We stooped again to our task. The rain beat on our arched backs without pity, soaking our hessian capes, finding weak points in the oilskins below, turning the soil around our boots into clinging mud. Only the thought of the alternative, splashing paint about in Aunt Kit's spotless kitchen, kept me going.

For the last three weeks it had rained heavily on and off after allowing us a good hay time once more, and a better than average harvest. As far as potatoes were concerned, we had been fairly well favoured for the earlies, and hit a good fortnight for the maincrop, just when the schoolchildren were turned loose for their potato-gathering holiday.

The boss had gone for a few rows of quality potatoes, Kerr's Pinks and Golden Wonder, but the bulk had been the heavy-cropping Arran Banner.

We had used the horses for spinning out and carting, as there seemed to be little advantage in using the tractor, and our pit-forming technique was designed around the box cart.

Since the start of the 'rainy season' the horses had come more into favour. A few more words of tractor terminology had come into our vocabulary, like 'wheel spin', 'bogged down', 'bellied out'. The boss had been heard to refer to the tractor as a 'fair-weather friend', which was hardly just.

At the same time he had decided on a two-row planter for next year, and had ideas of storing the potatoes in a converted building.

'We could use the trailer with side boards to haul the spuds home from the field. Then we could riddle on wet days inside. Anyway I'm fed up with sorting tatties outside in all weathers.'

The latter reason I felt would have been the weightier factor.

'Don't forget to put the shaws in heaps.' The boss's voice cut across my thoughts.

'Right boss. I keep forgetting you want to use them for cow feed; but won't they taint the milk?'

'Only if you feed too much of it. Anyway it saves getting into the marrowstem kale too soon, and I have to keep an eye on the cost of feed.'

A rise in the cost of concentrates had made the boss very conscious of a need to get as much milk as possible from home-grown feed.

'These turnip tops are in good condition, as yet no frost damage. The cows will eat them and it gives them a bit of green food.'

I knew the boss was intending to buy an electric fencer and strip graze the cows in the spring. 'I'm going to use a bit more nitrogen for early bite and try to get maintenance and two gallons from grass,' he had informed me.

At intervals during the day we talked of this and other allied subjects like silage and its place in the new scheme of things, but for the most part we were silent, trudging, chopping, rowing roots and heaping tops, two weary tortoises, brown and black and earthy, moving slowly across the land.

It was the middle of the afternoon, with the rain still finding gaps in our shields, when I moved into forbidden territory. 'There must be an answer,' I said.

'To what?' asked the boss.

'About the salt, you know, the mangels...'

'That's it,' he said. 'We're off home.'

'But it's hardly worth starting in the kitchen now,' I protested.

'You can wash off the ceiling,' he replied shortly.

'And what will you do?' I had a faint hope he might offer to help.

'I have some bookwork.'

It seemed there was always bookwork.

25
⤜ *The Drum* ⤚

'WHAT on earth is it?' the boss wanted to know. 'Here, you haven't joined the Salvation Army?' he went on anxiously. 'It looks like a drum.'

'It is a drum, at least a drum kit,' I replied self-consciously.

The object of our attention was a large cardboard box which had been delivered by the carrier. On the side was a diagram advertising the contents.

'The pipe band then. You're going to be the chap with the leopardskin, the drum major. Well, I never.'

'No, no, it's not that kind of drum. It's a jazz kit, you know, for a dance band.'

I could see he was nonplussed, so I explained. 'Some of us are getting together to start a band.'

'Ah,' he was stroking his chin and there was an amused twinkle in his eye. 'You and some of your smart-arse Young Farmer friends. A touch of the Joe Loss maybe.' He went off whistling a few bars of 'In the Mood', and when he had gone round the corner I heard him crack out laughing.

For the next few days the drum was the butt of his jokes. Every time our paths crossed he would be whistling 'In the Mood'. For Aunt Kit's entertainment he did an impersonation of Joe Loss conducting.

Despite a certain incongruity of appearance, I had to admit that he had the knee-bouncy action, the smiles and the nods off to a tee. At breakfast he did a drum solo using his knife and fork on any receptacle in sight.

I was getting a bit tired of it, especially as, having unpacked my gear, I was afraid I had invested in a sub-standard product.

After three days the boss remarked that he hadn't heard me rattle my drum, and I had to admit to its deficiencies.

'Bring it down and let's have a look. We might be able to put it right.'

Set up in the kitchen, the kit reflected the bargain-basement price I had paid for it. The big drum was uncovered at the front and balanced on small clip-on legs, giving it the appearance of an alarm clock. The legs fell off at the first thump on the foot pedal, which in turn produced a hollow echoing sound. The snare drum provided a tinny rattle like a child's toy. A single cymbal hung from a swan neck and a small tap box adorned the top of the big drum.

The boss took charge with enthusiasm. Several layers of rags damped down the foot pedal; a few copies of *The People's Friend* under the snare head made a world of difference. A solid wooden base for the big drum to rest on was planned, and Aunt Kit was enrolled to cover the front of the drum with a heavily starched linen cloth.

The boss decided that the large dangling cymbal was fine for gonging the end of a dance, but had limited use otherwise. A smaller, thinner version mounted on a spring would be the answer and add to the rather sparse furnishings. Sculls could be added at a later date; meantime the tap box would have to do.

The boss had gone out to look for an old inner tube when Aunt Kit produced the starched white linen. Stretched over the front of the drum, it could be mistaken for the real thing. The boss was impressed when he returned with a much-patched inner tube.

'Drummers bounce these days,' he said arranging the tube on a chair behind the drum. He tested the bounce, and picking up the sticks tried a tentative tap.

'Go on,' I urged. 'Let it rip.'

'No music,' he replied doubtfully.

'Sing then,' I said.

The boss needed little encouragement. Soon he was in full voice, rattling the snare head, bouncing enthusiastically, rolling his head, his foot pedalling the big drum. It was a compelling performance, if a little overloud and not very musical.

The cacophony brought Aunt Kit, tight-lipped, to the kitchen. 'Either it goes or I go,' she stated her terms firmly. 'I could go and stay with my sister.'

'The barn maybe?' The boss, a little shame-faced, was looking for a compromise.

'Somewhere right off the farm or else...' Aunt Kit apparently was in no mood for conciliation. She made her exit with an uncharacteristic bang of the door.

With Aunt Kit out of earshot I gave vent to a fit of the giggles. The boss's discomfiture had been so sudden and so complete, and Aunt Kit's ultimatum and exit so dramatic, that the boss could only sit there slack-jawed and at a loss for words.

Soon, though, I began to appreciate the gravity of my situation. I had nowhere to practise. Hoping for inspiration, I looked at the boss. He seemed to be recovering, and was thoughtfully stroking his chin with one of my wire brushes.

It was some minutes before he spoke. I could tell from the lightening of his expression that he had a plan; it was not often that he was completely outgunned.

'I've got it, he said finally. 'Young smart-arse has a nice barn loft, plenty of space, and the old folks have a bungalow away from the farm, and...'—he paused for effect before producing his trump card—'brother Tom, who is the real boss there now, is a first-class musician, used to play in dance bands all over.' He leaned back, beaming, and absentmindedly slapped the wire brush across the face of the snare head before hastily putting it down.

'You mean my mate Alan, I take it.' Since the boss referred to most of my friends by his bestowed title rather than name, I thought I had better be sure.

'Of course, and you know Tom well enough. He could be just the man you need, if he doesn't mind playing with a lot of smart-arse youngsters.'

I knew Tom, a quiet man much older than Alan, and now responsible for running the family farm. I had heard him play his violin at home, sitting by the kitchen stove. He had a velvet touch and was a real musician with dance band experience—what an asset he would be.

'I have another suggestion,' said the boss, commanding my attention more than ever. There're two chaps who play melodeons for the dances at Gartmore. I could have a word, if you like. Perhaps they would take you on. No pay, of course, but you would get some practice.'

He was as good as his word, and the following Saturday night found me trying to solve a transport problem.

Gartmore was served by a small private bus which operated infrequently, and certainly not in the early hours of the morning. There was nothing for it; I would have to transport my gear and myself on my motorbike.

The boss was there to see me off. I had improvised shoulder loops so that the big drum was mounted on my back, a reversed drum major position. Other items hung around me.

'You look like a mobile one-man band,' commented the boss, and as I set off I heard him call, 'Watch out for crosswinds.'

The dance was being held in the schoolroom. Self-consciously I entered with my bulky accoutrements, and running the gauntlet of curious stares from the waiting dancers, I made my way to the small platform in one corner.

My fellow musicians greeted me kindly. They wore clean work clothes with spotted hankies round their necks, and heavy boots of the sort favoured by shepherds.

Since they used these to beat out the time I had little difficulty getting the foot pedal action right. The dancing was mostly 'heavy', with only the occasional waltz, and in the rather small room, the war cries of the many dancers, the crash of the shepherds' boots and the considerable volume from the two melodeons allowed me cover to practise drum rolls and paradiddles without much critical notice being taken.

The dance had a starting time, but apparently only finished when people became tired. A hard core held out till four a.m., so I only had a few hours to wait after making the homeward trip before I could report the success of the venture to the boss.

On the third Saturday night, when the organiser came over to pay the musicians, he handed me a ten shilling note. I was thrilled beyond measure to receive this mark of my professionalism, and was quick to show the note to the boss the next morning.

He stood looking at it and fondling his chin. 'So you're a paid performer now?'

'Yes, sort of semi-professional, you might say,' I agreed expansively.

'Well now, as one semi-professional to another, if you get my meaning, there would be an introductory fee. Most agents would take ten per cent.'

I gave him a hard look. Was he joking? But he never joked about money. I decided to play along, and finding a shilling in my pocket, I ceremoniously handed it over.

With equal gravity he dropped it into a waistcoat pocket and went off whistling.

I waited for the sound of laughter as he went round the corner, but none came, only the sound of milk cans being rolled about, and the cheerful whistling of a man who has just done a deal.

As I say, the boss never joked about money.

26
⌾ *Downside* ⌾

SINCE my prankster mates Alisdair and Rob had moved on at the last May term, I had found myself more and more in the company of my Young Farmers' Club friends. Band practices were now a regular feature in Alan's barn loft. There were Alan and his older brother Tom playing violins; the local postman with his banjo; Bob, a competent accordionist; another accordionist who was learning and, although not much advanced beyond playing chords, made up the numbers; and myself, with a good bounce, a fair ear for timing, and getting better all the time at twirling a stick. My appearances with the Gartmore melodeon players had conferred a certain respectability on my drumming, but I was well aware that I had much to learn.

A slim, four-page tutor claimed it could make a drummer out of me in six easy lessons. Using this as a guide I assiduously practised the flam, the drag, the paradiddle and the roll. My 'mummy-daddies' had improved considerably playing for the reels and other set dances at Gartmore, but quicksteps tested me, and a tango was a near impossibility. I wasted time trying out gimmicks like bouncing a stick off the drum, catching it and giving it a twirl in my fingers without losing the rhythm.

A few of us turned out for a youth club party and felt we had acquitted ourselves rather well, although the exuberant youngsters, busy with their noisy cavorting activities, and getting our services free, were not very critical clients.

It was Tom who was approached for an engagement, a real dance, and a Hogmanay dance at that. The enquirer was, of course, aware of Tom's undoubted talent and past band experience, and totally unaware of how deficient the rest of us were in both departments.

Tom was full of misgivings. 'We have no amplification for a start,' he demurred, 'and some of you... well, you could do with a bit more practice.'

'We have a bit of time in hand,' said Alan reasonably.

'I have a friend who is a bit handy with electronics. I'm sure he could fix us up,' came the learner accordionist's contribution.

'I've had some experience, Tom, you know, at Gartmore, and I've heard the "Post" play his banjo in public.'

'You mean in a pub,' corrected Tom.

'Well, yes,' I had to agree. 'But he was near enough sober at the time.'

Tom sighed, and gave way under our blandishments. 'We need those amplifiers; don't forget,' he admonished the accordionist.

Time seemed to fly, for those of us in need of practice, and apparently for our electrician as well, because time passed and no electronic aids had appeared. His friend assured us that he was so handy that he was inundated with jobs to do for friends, but he had built the speakers and would borrow a microphone on the night.

This was hardly reassuring, but there was little we could do. Our practices were going well and we were growing in confidence, but on the night, nerves were on edge and self-esteem low.

We arrived at the venue early. There would only be a handful of people in the hall before the pubs turned out at nine o'clock, and we would be able to get our act together in the meantime. There was a collective sigh of relief that the sound man was already there and busy with an assortment of speakers, a skein of different coloured wires and, centre stage, the mark of a real band, a microphone.

There was a bar in the room, not yet busy, but, with a late licence, ready to fortify the revellers up to midnight. The banjo player made a purchase which he concealed in his banjo case after an initial sampling. It was clear his nerves were in no better shape than my own, but I felt a clear head was essential if I was going to give a virtuoso performance.

Tom placed me to one side of the stage near the front. The violins and banjo would play to the mike on the other side away from my noise, the accordions to the back and central.

We played a waltz while our handyman was putting the final touches to the speaker arrangement. In the quiet hall the violins were sweet and melodic, the accordions blended well, the banjo twanged, and I provided a well-damped foot pedal, a smoochy rotating right brush, and a left brush that flickered from the snare head to the cymbals, catching the light with its exaggerated movements. My bounce, I was sure, was slow and sexy; I was well pleased with my performance.

A couple danced to our music—they could have been the bar attendants—and clapped enthusiastically when I brought the dance to an end with a gong, a roll on the drum, then another cymbal clash.

We played another number, a rousing quickstep, and I managed a successful toss of a stick with a catch and twirl and was beginning to feel less edgy, when out of the tail of my eye I glimpsed our electronics 'expert' standing with some wires in one hand and a screwdriver in the other. On his face was an expression of puzzlement. I caught Tom's eye and brought the dance to an end. 'I'm not too sure which way round the wires go,' said the hapless individual.

'But you've done it before; you're supposed to be an expert,' challenged Tom, now seriously concerned.

He shook his head. 'Well, no, not really. I made up the speakers myself; I thought the rest would be easy.'

'Well, it's got to be trial and error,' said Tom. 'Just get on with it.'

Throughout our next few numbers, a variety of sounds came through the microphone as our expert touched different combinations of wires.

At last, just before the pub customers burst through the doors, the electrician got it right and we were at full volume, the microphone raising the sound of the strings to compete with the more aggressive instruments. It was a very acceptable sound and soon the floor was full of rather merry dancers.

All was going well and we had accepted a number of drinks handed up by hospitable people, which if anything improved our performance, but it was too good to last.

At eleven o'clock an amplifier developed a fault and started to drone. Soon after, the mike was in trouble, causing the violin music to distort. This was closely followed by a rush on stage by the electronics man, who secured the situation by ripping out the wires.

Without amplification the violins and banjo were lost in the noise from the floor, and the accordions were struggling, but we soldiered on. Luckily more people were drinking than dancing.

One tall, willowy blonde girl, who was obviously a good dancer and was always on the floor, had a succession of partners. I began to think maybe she fancied me, as she kept looking at me. I gave her a brilliant Joe Loss-type smile, then I lowered an eyelid and gave her a meaningful look.

The next time round she stopped. I leaned down from the stage and her partner stood by a little embarrassed. She reached up on tiptoe, her lips close to my ear. 'Your band's lousy,' she said, and danced on.

I looked around the band. Perhaps she was right. The banjo player had given up and, having imbibed freely from his banjo case as well as the hospitality from the floor, was slumped, and out of action. Tom and Alan were sawing away apparently soundlessly; the chord playing accordionist had given up and was at the bar drowning his sorrows; the other accordionist and myself were the only ones capable of making any impact. I had just had my self-esteem totally destroyed. There was only half an hour to midnight, so as a man we packed away our gear and headed for the bar. As Alan said, putting our situation in a nutshell, 'If you can't beat them, join them, especially on Hogmanay.'

Tom refused to play with us any more. Without him we were rudderless, and our musical aspirations took a downturn. I advertised my improved drum kit in the local paper.

I made a small profit on the deal which I didn't talk about, just in case the boss still considered himself my agent and claimed his ten per cent. I knew I should have fired him after the Gartmore business.

Inevitably I was to lose touch with Alisdair and Rob. I had missed them when they left, but soon got over it. Perhaps I had outgrown the pranks we used to perpetrate together, and I had turned my back on the nomadic journeyings of my earlier years in favour of a live-in, as family, steady job.

I often laughed at the memory of 'typicals' struggling into work clothes at two o'clock on a Sunday morning; it had been one of our funniest. And here I was in the early stages of becoming a 'typical'.

Roots were being put down, loyalties were forming, friendships, activities, the things that keep a man in one place, and yet I knew that one day I must move on. The leaving would not be easy, but I had to make a life for myself.

I wanted to progress to management. Almost certainly this would have to be down south, and I knew that first I would need some formal training. I must find a way to get to Agricultural College; I couldn't see a way, but I would keep on looking.

I located Alisdair some years on in the Blackford area. He had gone back to bothy life after a failed marriage. I was there for the games and he was able to offer me a bed for the night. He was on a self-catering arrangement and in the morning went round the stackyard to find our breakfast eggs.

Apparently he kept a sharp eye on hens which laid away and regarded this as a legitimate perk. The occasion is well flagged in my mind, as, for the first and only time in my life, I confronted a goose egg for breakfast.

It occurred to me that Alisdair was showing symptoms of the 'typical' bothy dweller, but I don't remember whether he slept in long-johns and socks.

27
⟫ *Blackie* ⟪

I HAD christened Fanny's foal Blackie for obvious reasons.
We had been able to let him suckle for five months, but the
weaning cries were now in the past, and Fanny, with a bit of
luck, was back in foal again.

The stallion leader had been nice as pie this time. Perhaps
because he had spotted the pail of warm water ready by the
kitchen door before the action commenced, but more, I think,
because Fanny had held first time.

'That's a fine little mare you've got there...' He hesitated,
and I felt sure he was about to say 'boy' but instead he said
'young man'. 'Yes,' he went on. 'A hot little property. A real
breeder. Not all the mares hold first time, some are a bluidy
nuisance, three or four times some of them. Some are a dead
loss.'

He was full of praise for the foal. 'Like his dad; that one
could win you a prize at Stirling. Is he well handled?'

I assured him that this was the case. I had accustomed
him to the halter while running beside his mother. I led him
frequently, walking and trotting. I was forever picking up his
feet, paring them to keep his frogs on the ground. I had looked
in his mouth so often I felt sure he would almost open on
command. We were the best of friends. I knew I would miss
him when he went.

The boss had decided to sell Blackie as a rising one-year-
old rather than as a two-year-old or older. He had explained

his thinking. 'Look at it this way. We might have another foal at the end of next summer. If we kept Blackie we would have five horses on the place. It would begin to look like a ranch. Anyway, we have a three-stall stable. Of course we could always knacker Blossom,' he added brightly.

I knew he was joking about Blossom, but he had a point. If we kept young horses on, we might be tempted to keep one and break it in to replace an old stager. Blackie would have to go.

The stallion leader's goodwill was boundless. 'When you want to get the foal ready for the show, I could fetch somebody who knows all the tricks of the trade, that is, unless you've done it before.'

I had to confess I had not the slightest idea how to prepare a foal for the show ring. 'I'll have to talk to the boss,' I said.

—

On March 12th, the evening before the show, a small elderly man arrived and we washed Blackie's feet in suds. He cooperated fully, standing quietly as one foot at a time was placed in a small zinc bath of soapy water, washed, rinsed, and finally dried off with wood flour.

'We'll tart him up at the show,' the man said. 'Just give him a load of clean bedding, wash off any soiled bits in the morning and don't let him lie down in the lorry. I'll see you there.'

By the time I had to parade him, Blackie was resplendent in ribbon and raffia. I had watched closely as the expert dusted powdered resin round the hair covering the hoof tops and then wiped under the hair with a bar of soap. A final light brush out, Blackie was ready to show his paces.

'Makes his feet look bigger,' confided my mentor.

As if aware that everything had been done to accentuate his good points, Blackie paraded perfectly, walking, trotting with a showy flurry of white leg hair, head held proudly high, and with it, a stock-still stance when the judge wanted to look him over. When he ran a hand down a fore leg, Blackie smartly raised a hoof. 'There won't be any trouble shoeing this one,' the judge said to a steward. I felt we had scored a point, and was a little disappointed that the judge hadn't wanted to inspect the mouth.

In the event we collected the third-prize rosette, and I paraded Blackie to some keen bidding at the sale. I could, of course, have been mistaken, but I felt sure I had seen the stallion leader in the crowd. I was equally certain that I had seen his eyelid flicker in the direction of the auctioneer, just when the bidding seemed to have tailed off on a couple of occasions.

I stayed with Blackie until his new owner led him off to a waiting lorry. The excitement was over, and I had a lump in my throat as big as a rock. I repaired to the pub, ordered a pint and, not feeling in any mood to socialise, found a seat in a remote corner.

The boss came in. I kept my head down. He would be puffed up with the money Blackie had earned him; I could only nurse my sense of loss. He was the last person I wanted to talk to at the moment.

Soon the purchaser came in and went straight up to the boss.

They shook hands, all smiles and goodwill, and talked for a time over their pints. Then they shook hands again, and the purchaser turned to leave, but apparently having just remembered something, turned again towards the boss. I saw some money change hands, but the purchaser was shaking his head. There was more dialogue, then more money changed hands, and after more effusive hand-shaking and back-slapping they parted. By this time only the new owner of Blackie was smiling. Having witnessed the boss part with a 'luck penny', I continued my brooding vigil. If the beer was doing nothing for the lump in my throat, then the expression on the boss's face as he was accosted next by the expert of the ribbon and the raffia did. This time there would be an agreed fee to hand over. The man generously bought the boss a pint out of the cash received.

It was when the stallion leader sidled up to receive a surreptitious payment that I realised there was more to selling a foal than just running it up and down.

The lump in my throat was not getting any better. I could see from the slump of the boss's shoulders that parting with hard cash had depressed him; I went over.

He glowered. 'You're not after money as well?'

'I thought I might buy you a pint,' I said.

'That's more like it,' he growled, then, looking at me keenly he suddenly smiled. 'You must be missing that bit foal, I reckon. Tell you what, I'll buy you one.'

28
❧ Scholarship ❧

*D*RYING winds in early March allowed us to get the grassland fertilised for early bite. Donald in the shafts of the fertiliser distributer hardly left a hoof mark on the surface. Time and again during wet spells the horses had proved their worth, but equally we sometimes wondered how we had managed before we had the tractor.

The fact that our neighbour had purchased a buckrake and was set to make grass silage later on had the boss thinking. He would have liked to buy a few more tractor implements, but felt he had to justify his expenditure with more output.

As usual he used me as a sounding board.

'First we have to look at what we can't do. We can't keep any more cows; the byre's full.' He paused to ponder his next pronouncement.

'You could change to black and whites, get more milk from the same number of cows.'

'They're bigger; they eat more.'

'Even so, the bull calves are worth a lot more.' I carried the argument to its ultimate, even though we'd been over the ground before. I knew the boss was half inclined to make a gradual shift to Friesians, but was constrained by a long family tradition of being 'Ayrshire' people.

'It's not as if you were all "Ayrshire" anyway,' I said. 'You have a few Shorthorn crosses, that old kicker for instance.'

'Well, I don't say you haven't got a point, but that's something for the future. Looking at what we've got now, I think we have to get more from our grazing land and try to squeeze in a cash crop.'

'What do your friends down south do?' I had learned that the boss was preoccupied with things south of the border, because he was still in touch with members of a family who had migrated lock, stock and barrel by train, and were now well established and prospering in the better weather of southern climes.

'Well, now,' said the boss, 'they would probably bang some winter wheat in after the tatties. We could do that of course and hold the oat sowing and the undersowing over till the following spring. It would leave us short of grass at some point, though.'

'Direct reseed after the wheat?' I wondered.

'We might miss the crop of oats; no, we must manage our grass better; a balanced fertiliser, and Nitro Chalk and strip grazing.'

'And silage?' I wanted to know.

'We'll look over the hedge at our neighbour; let him make the mistakes first.'

'Then there's Aunt Kit's hens,' I said. 'I was talking to Peg about them last time she was round. She reckons those old Wyandottes, Barnevelders, Light Sussex and such-like, spend half their lives broody, laying away, or underneath the cockerel enjoying themselves.'

The boss chuckled. 'That sounds like Peg. What do we do about it? Did she say?'

'Get some of those hybrids that are about now. Put them inside on deep litter. Give them some extra light and watch the money roll in.'

'She didn't happen to say where we would keep them, exactly?' queried the boss.

'No, not exactly. She said there must be plenty of secondhand buildings going cheap, ex-army huts and so on. She said we should keep an eye on the ads in the papers.'

'Then you're just the chap,' said the boss. 'Just think what a bargain you got out of the paper with that drum. Did you lose much when you sold it?'

My mumble was noncommittal. There seemed to be little mileage in admitting I had made a profit.

—

We settled in our places at the Young Farmers' Club. There was a general air of anticipation. The elderly gentleman out front with Mr Jessiman and our chairman was a governor of the West of Scotland Agricultural College. He had a 'Dr' before his name, and his subject was 'Grass, the Neglected Crop'.

I had known for a week that I was to propose the vote of thanks, and my stomach was revolving with nerves. I had determined that this would be the best 'impromptu' speech ever.

First I had written the speech, then reduced it to headings on a postcard. It was my intention to speak without notes, and I hoped my inner tremors would not transfer to my voice when the time came.

I had a solid week of rehearsal behind me. I had declaimed to the swaying buttocks of the horses, the exhaust pipe of the tractor, a horse's collar in the stable, the open spaces of the moor. I was word-perfect.

It was the last lecture of the season. I could see the boss was enjoying it by the number of times he wagged his head in agreement. Sometimes he and Peg would whisper together.

I hoped she was telling him about the bad behaviour of the farmyard hens.

It was a masterly lecture, and as always I became totally absorbed so that it came as something of a surprise to me to be summarily called to make my speech.

I stood up, and for a long moment said nothing, not because it was good technique, a mark of confidence, or because I was waiting for complete silence. It was just that I had forgotten what I had intended to say.

The chairman looked at me enquiringly just as my memory returned, and I was soon in full flow, my nervousness gone, my articulation perfect, as unselfconscious as when I had delivered my speech to Blossom's rear end in the stable.

I likened the speaker to a good cook who had to bring with him all the ingredients. The ingredients I listed were highly flattering, and I was able to assure him that the resultant mix had been a great success.

When I sat down, a storm of clapping and foot stomping broke out. I hoped some of it was for the excellence of my oratory.

The meeting was breaking up. Someone nudged me. 'Jessiman wants you.'

Mr Jessiman was indeed pointing at me with the stem of his pipe. He was with the speaker, and they greeted me with great friendliness. The lecture had been well received and my vote of thanks and the storm of applause had obviously gratified him.

'Have you ever thought of going to college?' he asked.

I assured him that I thought of little else.

'There's a bursary,' he went on, 'open to sons and daughters of agricultural workers or to bonafide agricultural workers.'

I told him I was eligible on both counts, for although my father had spent his early life around horses in 'big house' service, and was now a roadman, there had been spells of farm work sandwiched in between.

Excitement had played havoc with my diction. I babbled my thanks more than once. He would send details in the post. No, I needn't expect it the day after tomorrow, it would take a little time, perhaps a few weeks.

I was vaguely aware that the distinguished gentleman who regularly occupied a seat at the front, had been observing the proceedings. Over the months, he had taken to having a chat with me, asking about club competitions, my prospects, my aspirations; he seemed to have an interest in me. He was an important man, with a title, land and a big house. He always tried to put me at my ease; even so, I was still a little in awe of the tall aristocratic figure with the Duke of Wellington nose and the upper-crust accent.

He was waiting for me by the door; his hands were in his jacket pockets, thumbs exposed at the front. He had obviously been aware of my chat with the speaker.

'There's a bursary you can apply for. It can be funded by the Department of Agriculture or the local Education Authority. I, of course, could look out for it if you decide on the latter.'

I knew that 'Sir' had a finger in many pies, from milk marketing to local government.

'I'll send details in the post.'

'It might be a few weeks, I suppose?'

He smiled. 'I'll try to do better than that. I expect you're in a hurry.'

I thanked him, and as I turned to go, he said, 'And by the way, congratulations, I understand you will be appearing at the Kelvin Hall in our quiz team. The finals, no less. Good luck.'

I thanked him again. It was a timely reminder.

My excitement knew no bounds. For the next few days I babbled about the bursary and little else. Would I be accepted? Would 'Sir' or the 'Dr' remember to send the forms? Should I make a direct approach if nothing came soon?

On the fifth day, when desperation was setting in, I missed the postman, but Aunt Kit came to find me bearing a large buff envelope. 'Sir' had been as good as his word. The very next day the same forms came from the 'Dr'.

'Now we might get some peace, Aunt Kit,' said the boss.

'He might need a bit of help with the forms,' she suggested.

'Well, he's got one lot to practise on and one lot for best,' said the boss.

'You'll have to look out for a new lad,' I said soberly, the thought of leaving what had become my home tempering my excitement.

The boss nodded. 'But not right away. There's a lot we have to get done before then. You could start by telling Aunt Kit what Peg said about the hens.'

'You mean exactly?'

'Well, maybe not exactly,' said the boss.

———

The national finals of the Young Farmers' Club quiz competition were only weeks away; I had to apply myself to study. We had done well to get this far. There were five in the team, three young men and two girls. The questions were to be equally divided between farming and general knowledge.

Our strategy in previous rounds had been for each to mug up on a branch of farming so as to cover the field. The general knowledge area had to depend on luck.

On this occasion the two other lads would do crops, seeds and fertilisers, the two girls, pigs, poultry and cows, and I was to do horses only. Rumour had it that as horses had not featured previously, they would come into the reckoning for the final.

My recently acquired *Thompson's Elementary Veterinary Science* provided diagrams of the skeleton, the muscular system, the respiratory system and a list of ailments, most of which I had been unaware of.

I wrote things down and used the horses as models. Blossom, whose bones were nearer the skin than Donald's, was my skeleton. I would run my hands over her, muttering 'cervical vertebrae, dorsal vertebrae, stifle joint, tibia...' I was soon well versed from the poll to the pedal bone.

Donald was my ailing patient. I would locate the site of the ailments and pronounce the names—curb, thorough-pin, splint, sidebone, laminitis...

The blacksmith's son pressed a book on me which dealt in depth with the foot and shoeing. I learned about palisade layers, horny tissue and where not to drive a nail.

By the time the quiz came round I was well primed.

In the event, there was only one question on the horse: 'Where is the frog and what is its function?' I thought even the boss could have answered that one.

We did not win or even make the first three. Disconsolate, the other two male contestants sat in the front of the car going home. The two girls, equally subdued, shared the back seat with me.

I felt I had a duty to cheer them up. Before we had cleared Glasgow in our trundle homeward through the starlit night, the merriment in the back seat was in stark contrast to the gloom in front.

I suppose that, in the dark, the back seat of a car is as good a place as any to enjoy being young, carefree and just a bit silly.

29
⇌ *Shearing* ⇐

'WHAT do you know about sheep?' The boss's question caught me unawares. I had been thinking of other things, like when I might expect to hear news of my bursary application, whether I would like living in town, whether I could live on two pounds a week. This was the sum total of the grant, and out of this, I must endeavour to keep body and soul together, and buy books.

'You can't just keep thinking about that scholarship all the time. You must keep on learning,' the boss went on, as if divining my thoughts. 'The letter will come in its own time.'

He was right. I had to get on with things, so I switched my mind to sheep. 'All I know is they get tangled up in brambles and buried in snow. They seem to be bent on causing mischief all the time.'

'Perhaps you realise,' said the boss heavily, 'that they have to be clipped once a year.'

I admitted to that amount of knowledge.

'Well, it's like this. I don't want those smart-arse college chaps to think your practical training has been neglected, so...' He paused to lend weight to his punch line.

'So you've arranged for me to help old man Duncan with his shearing.'

The boss was clearly taken aback. 'But how...?'

'Heard it in the pub. Mr Duncan had managed down for supplies.'

'Then you'll know when he expects to start?'

'The first or second week in July. We have to watch for the smoke signals.'

'He's not that far out, and they do have a telephone.'

'Far enough by all accounts. You might have to draw me a map.'

'No need for that; the road doesn't go anywhere else.' The boss was smiling and had that look on his face that said he had saved the best bit for last. 'By the way, it's oil lamps and hand shearing. They don't have electricity in the steading.'

The boss was in good spirits. The early bite had responded to the two hundredweights of Nitro Chalk per acre which had gone on in March, although some 'poaching' weather had kept the cows off it for two weeks longer than he had planned.

He had been sympathetic when my old BSA had started playing up earlier in the year, and had found someone to patch it up for me.

'Better look in the papers, find another one, and then sell the old thing. We could always block a hole in a hedge with it if nobody wants it.'

But someone did want it. Once again advertising brought me a customer and a small profit. Bikes were still scarce, and I had to pay more than I had intended for the 1935 Velocette (push rods, overhead cams) with which I replaced it.

It was while perusing the newspaper ads during this period that I spotted 'Secondhand shed for quick sale, dismantled ready to erect...' in the miscellaneous-items-for-sale column.

The dimensions seemed right for a small deep-litter house, and it was on a smallholding within reasonable transporting distance. Ex-army, it had apparently never been erected on the farm and was clean as far as livestock was concerned.

The boss was interested. Peg had been round and talked to Aunt Kit about it. Aunt Kit was now convinced that it was unfair to expect her to climb into hay lofts, under stack prop wigwams and along the backs of hedges looking for laid-away eggs. And it seems she had noticed these old heavy hens spent half their lives 'clocking'—she made no mention of any other misdemeanour—and it was time the boss made things easier for her.

We were pleased to get a yes vote from Aunt Kit, and made arrangements to see the building one evening.

The smallholding seemed to be extremely well stocked with a mix of livestock. The owner turned out to be a tall, droopy man with a hangdog expression.

'Of course I would have had it up by now and full of calves if the wife hadn't been taken ill. I haven't been too good myself, and my son's worse than useless; he's got a drink problem. Still mustn't grumble. I'm prepared to sell it for half what it's worth just to get some cash in the bank.'

'Show it to my foreman,' said the boss. 'I'll just go behind the building over there. I'm bursting for a pee.'

The shed was in good order and someone had numbered the sections as it was being taken down. Some new bolts, maybe a roll of roofing felt to patch up the roof panels; it looked a good buy.

When the boss came back the smallholder was ready to shake on it, or perhaps he was holding out his hand expecting the boss to drop a cheque into it.

Instead the boss said, 'I'll give you half what you're asking, otherwise no deal.' The man haggled, and we were in the car when he finally succumbed.

The boss dropped a cheque, already written, into his hand and got the man to sign a receipt, also prepared in advance.

I was amazed.

'The man's a dealer,' the boss explained. 'I could tell by the stock. Too many to stay more than a few days between markets, and when I went behind the shed I had a nose round. A healthy-looking young man was washing out a lorry, and a woman who might have been an Amazon was working in the garden. He would have picked up the shed for next to nothing. All he wanted was a quick turnover. That's what dealing's all about.'

I looked at the boss in admiration. 'I see I have a lot to learn.'

He grinned. 'You can start with the shearing.'

'I thought hand-shears were for trimming round corn stacks,' I grumbled.

'Then there's this building to put together. A bit wall for it to sit on, plumbing, lights. You'll be able to do all that.'

'More learning,' I said. 'That's enough to be going on with.'

———

The Duncans' shearing fell neatly into that gap between piking the hay and getting it in. It was the start of the second week in July, and a fine day, as I left Gartmore behind and followed a minor road which eventually led me to the loaning over the moor to the Duncans' homestead. No more than two wheel tracks, with grass growing in the middle, and seemingly endless; there was no habitation in sight. I didn't mind. I was enjoying the feel of the more powerful bike beneath me as I dodged the potholes. The sun was warm on my back, and I was heading for a completely new experience.

The boss had said, 'Of course, if we're busy here you can

maybe go only for the afternoons and evenings... or just evenings even.'

As things turned out, all I was missing was a bit of white-washing. The boss had made a rather reluctant start on the byre as I left.

The track was rising gradually, and despite the bright sunshine I was aware of a chill wind. Bushes and occasional small trees leaned away from the prevailing wind, stooped and twisted like very old people in a harsh environment. I could understand why the sheep kept their fleeces until July. It was hard country.

Finally, after what seemed an age, the road levelled and ahead of me, nestling in a hollow, low-roofed and stone-built, was the steading. It blended perfectly in the landscape, snug below the wind and tidy.

As I approached the house I passed the sheep, about two hundred Blackface, enclosed by stone walls. A Garron pony stood, sturdy in old age, untethered, a well-used saddle and bridle on a bench nearby. When three black and white collies burst barking from a building, I could see exactly how the old man did his gathering. A long, low whistle caught up with the dogs, and they flattened, observing my passing with suspicious eyes.

A few sheep were held behind hurdles by a barn. Inside I found Mr Duncan. He was using an old-fashioned clipping stool. He sat astride one end, a sheep cradled to his waistcoat, and the shears in his gnarled hand snipping busily. He didn't speak till he had finished his sheep and pushed it out through the back door of the barn.

'Get a sheep,' he said.

I went to the pen and grabbed the first pair of horns that got in my way.

'I have only one stool, so you'll have to work on this sheet; here, I'll show you. First sit it up on its arse, clean the front, then down the shoulder...' He demonstrated. 'You've never done it before?'

'No, but I've seen sheep shearing at shows.'

'Aye well, they shear; hereabouts we clip.'

We went to the pen and got ourselves a sheep apiece. Mr Duncan settled on his bench and I was on my own.

My sheep seemed to have a mind to struggle. I was glad it wasn't a Border Leicester. Mr Duncan's sheep seemed to be happy to share the rough comfort of the shepherd's clipping stool.

I was still going down the first side, and trying to remember what he had said about putting my fist somewhere to straighten the back leg, when the old man released his sheep.

He made no attempt at conversation, apart from the odd grunted comment like, 'Don't pull the wool or you'll nick the skin. There's some yellow stuff there if you do.'

After two hours, with my back hurting and my fingers weary from squeezing against the tension of the shears, he said quite suddenly, 'Your boss used to play for the Rovers.'

I waited, certain there was more to come. I straightened. He motioned that I should keep clipping. About twenty minutes later, when I was nearly dropping from backache and fatigue, he said, 'Dinner.'

As we walked to the house he said, 'Big chap, hard to pass. If somebody did manage to run round him, he would click his heels.' He laughed shortly, but said no more.

I thought he must be nearer eighty than seventy. The long angular frame was bent from a lifetime of toil, but like the trees and bushes I had seen on the way in, was firmly rooted in the inhospitable moorland. The corded arms were apparently tireless, and the habit of conversation had been lost in the solitary lifestyle. Athough his step was faltering, on the clipping stool he seemed indestructible.

I wondered if the boss had known what he was letting me in for. I suspected he did. Mrs Duncan was a female copy of her husband, except in one respect: she was as garrulous as he was taciturn.

After a meal of lentil soup and home-baked scones, Mr Duncan moved over to an old armchair, where he stretched out and went to sleep. I was pleased to observe this, as I needed a little time to rest my aching back, but after half an hour of close interrogation by his spouse, I was glad when he twitched and jerked back to wakefulness. Immediately, he got up and made for the door. 'Time to get started,' he informed me.

Just before tea and scones arrived at four o'clock, his short barking laugh broke the silence. 'Didn't often get seen by the referee either. Oh aye, he was a hard man to get past.'

At eight o'clock he spoke again. 'It's a bit dark in here now. Not worth lighting a lamp. Might as well stop.'

I agreed with alacrity. It was indeed gloomy in the old barn. My back was killing me, my right hand likewise, I could have eaten a horse and I had been starved of conversation.

Mrs Duncan plied me with bacon and egg and fried potatoes and questions, followed by a generous helping of rice pudding and more questions. Afterwards she insisted I rest in an armchair and tell her all about Aunt Kit, the boss and life outside the moor.

Well fortified, with the work behind me, I was preparing to enjoy the ride home, but with night descending the moor was full of dark peat bogs and ghostly shapes. It took a special kind of person to live with it; I felt the old couple I had just left were a bit special.

The next morning Mr Duncan greeted me with the words, 'We should break the back of it today.' I didn't think the phrase was too apt, considering the condition of my back, but we soldiered on and indeed had nearly finished by nightfall.

Conversation had again been in short supply. At one stage the old man had laughed and said, 'I could tell you a thing or two about your boss when he was a young 'un.' I hoped he would, but he didn't elaborate.

As I was about to take my leave, he suddenly held out a rough hand. I was aware of the strength of the fingers as we shook hands. 'Thanks,' he said quite simply, but I felt there was more to come. I waited. 'It's nice to have somebody to talk to,' he said. 'Come and see us when you like.'

As I rode across the moor, my headlamp picking up the grassy middle of the road, the potholes and now and then the silhouette of a wind-ravaged bush, I was filled with wonder at the persistence and hardiness of the people and their livestock, the vegetation and wildlife that survived in such a place. I felt I would like to get to know the moor and the people on it a lot better. Come and see us when you like, Mr Duncan had said.

I made up my mind that I would.

30
⇜ *The Rovers* ⇝

*T*HE site for the deep-litter house had been selected—fairly level, near water and electricity supplies, handy for Aunt Kit to get to—and the footings had been dug.

It was to be an 'in between' job, but the boss's enthusiasm knew no bounds, and we spent as much time as possible on it, giving way only to the demands of haymaking, horse-hoeing and little else.

The boss was very much in charge, having watched the builders erect our draff silo, but I thought it was a bit ostentatious to keep referring to the peg in the middle of the site as the datum point and the corners as quoins, particularly as we were only going two blocks high.

Once the dwarf wall was built, the sections went together quickly, and with the shell up we began to plan the inside.

The boss had decided to have an egg room in the centre section of the shed, with a bank of communal nest boxes forming a wall either side. Aunt Kit would then be able to collect the eggs through hinged flaps opening into the egg room and pack directly into the boxes for sale.

The boss was rather pleased with his idea. 'Aunt Kit won't need to get her shoes dirty; she'll never have to go in with the hens.'

'What about feeding and watering?' I wanted to know.

'Water will be on a ball valve, and feeding... well, I'm sure you won't mind obliging Aunt Kit.'

I persisted. 'Neither of us are carpenters. Who's going to build these nest boxes, extra doors and such like?'

The boss's planning clearly hadn't gone that far, and he stood thinking for a time before he said, 'A mate of mine. He's a carpenter and undertaker.'

'You mean that grey-haired chap who plays centre half for the Rovers?'

'That's him. He may be getting a bit grey, but he's only about thirty-nine, and he's still better than any of the youngsters.'

'There's quite a few veterans like him in the team, still too good to give way to the younger chaps. You used to play right back, didn't you?' I quizzed the boss.

'A long time ago. The farming had to come first when I was left on my own. You play at right back for the Young Farmers sometimes. Are you any good?'

'Well, folk say I'm hard to get round. If anybody does pass me, I click their heels.'

He gave me a hard look, but said nothing.

—

The rapid progress on the new enterprise had prompted Aunt Kit to start depopulating the errant farmyard hens. These were increasingly on the supper menu. Boiled, then roasted to a palatable golden brown, they would have been quite appetising but for the boss's tendency to personalise the dish: 'Is this the old Barnevelder with the bumble foot, or the Wyandotte that was scouring?' And Aunt Kit would answer in kind, 'No, it's the old Rhode Island that got a bit crop bound,' and so it went on.

When Peg came to supper it was worse, because, with her undoubted expertise, she was in a position to explain to me the correct way to wring a hen's neck. This would be compared with the ritual murder of other classes of fowl, with the attendant difficulties when it came to ducks and geese and large things like turkeys.

Catching the free-range hen was also, it seemed, an art. You chased it around making excited cockerel sounds until it squatted to oblige, at which point you picked it up and rapidly snapped its neck. This one filled me with indignation. It was unfair, unsporting. I started to challenge her with 'How would you like...' but with the boss's eye on me I desisted.

After a time even the boss seemed to tire of the boilers, and when I brought news from the pub that a local poultry dealer would pay eight shillings for a good heavy bird, he was quickly on the phone and my travail was over.

—

The carpenter arrived to supply the necessary skills for the construction of the egg room, and it was the boss who brought up the subject of football. 'This young man would like a game with the Rovers,' he explained. 'He's a right back, hard to pass and a heel clicker.'

'Sounds a bit like you used to be,' said the carpenter. 'But if you recommend him, he might fill in for old Tommy. He's not a youngster any more and is carrying an injury. Anyway, we're at Gartmore on Saturday night, and Tommy's not too keen on playing up there. If you remember he had a spot of bother when we played them last time. I'll talk to the others.' A phone call on the Friday night confirmed that I was in the team.

The Village Summer League was a keenly contested affair, and more than a little partisan. The standard of play was high with a sprinkling of ex-professionals and many who had had experience of junior football or better. I was aware of the honour bestowed, even if it was stop-gap and a one-off experience. I wondered how much my selection was due to the boss's recommendation; as far as I knew, he had only seen me play on one occasion.

I knew a few of the players and was greeted civilly. The carpenter said, 'I hope you're as good as your boss says,' which left me wondering just what he had said. The opposing manager looked into the dressing room to say he hoped for a good game and no rough stuff. He was well on the way to being drunk, and seemed ready to be disorderly on his own account.

'Village bobby,' explained one of my fellow players. 'He gets a bit over-excited on match nights. He'll be on and off the field like a yo-yo if any of his players take a knock.'

'Is it going to be a rough game?' I was beginning to read the signs.

'Could be. The crowd don't like us up here either; they think we're a bit too fancy,' replied my informant. 'You'll be up against a sturdy outside left, comes at you like a tank. Are you as strong a tackler as your boss says?'

'What did he say?'

'Well, the word is, nobody gets past you; if they do, you bring them down. It seems your boss thinks you will be as good as he was some day.'

The carpenter joined in. 'I played beside your boss. He had a bit of a reputation, but he didn't deserve it. It was just that he never let up; even lying on the ground he would try to stop his man. It might have been he touched somebody's heels together now and then, but he was always ready to pick the chap up and apologise to the referee.'

I began to see what was expected of me. We trotted out to catcalls from a crowd of perhaps two hundred. The game started, and a foul at the far end had the Gartmore manager on the field gesticulating and threatening and the crowd baying angrily.

While the local bobby was being dragged from the scene, my carpenter friend came over. 'The ref won't be on our side. He'll want to get to his car in one piece at the end of the match,' he said. 'Just try not to get involved in any punch-ups; tackle hard and fair and you should be all right,'

The first half went well for me. Being tall I could get to high balls; I was fit from my running and could cover a lot of ground. I had so far avoided bodily contact with the outside left, and managed to stop most of his tank-like runs by tapping the ball into touch. The crowd seemed to approve of my no-nonsense style. I heard one old man say to another, 'None o' that buggering about with fancy footwork. He's more like a Gartmore player.'

At half-time the captain clapped me on the shoulder and said, 'Well done, lad, that's just what we wanted. No fancy stuff, just solid defending and no trouble.'

'I'm not much of a ball juggler,' I confessed, 'but so far I've managed to stop the outside left from getting many crosses in.'

'Keep it up,' said the captain. 'You're doing fine.'

Early in the second half the Rovers scored to go one up. The Gartmore manager disputed the decision, claiming offside, and had the referee by the lapels before he was dragged off, screaming threats. The crowd bayed its displeasure and thereafter became more hostile.

It was in this atmosphere that I blotted my copybook. The winger was on a run down the touchline. I was a stride short of simple tap into touch, when I leapt and slid. He punted the ball ahead and hurdled my still-sliding torso. A flailing hand must have touched his heels together because he crashed headlong to the turf.

I was quickly up, and was bent over the prone body when something crashed sickeningly on the back of my head. I sank to my knees beside the other casualty and a blast of colours ranged in front of my eyes before my head cleared and I could look round.

Expecting to be confronted by a macho male in a Kung Fu stance, I was surprised to see instead an old lady. She made the slightest gesture with an umbrella she held in her hand, rather like the minimal nod of a farmer bidding in the mart. The message was unmistakable.

'It's his ma,' somebody whispered. 'She'll have you every chance she gets now. Nobody hurts her boy and gets away with it.'

The referee gave a foul against me but, perhaps thinking I had been punished enough, took no further action.

The game proceeded, and inevitably their outside left collected a pass right on the touchline within feet of his formidable mother. Instead of the uncompromising tackle I stood off, feinting this way and that, but the winger didn't advance. Instead, with time on his side he crossed accurately, and a goal resulted.

There was some question of offside, but this was not hotly contested, and the referee had assured himself of a safe passage to his car at the end of the match.

A few Rovers supporters directed their anger against me, advising me to cut out the fancy stuff and get stuck in, but there were no recriminations from my team-mates, and I had gained a few supporters in the ranks of the Gartmore spectators.

As we trooped off at the final whistle, the captain said to me, 'Well done, lad. A small misjudgement in the second half. Maybe you should have upended that chap before he got his cross in, but never mind, one each is a fair result. We won't get abused by the crowd and we can all drink in the same pub.'

I was surprised to see the boss in the pub.

'Got there in time to see the second half,' he said. He ordered a pint for me.

'I played better in the first half,' I said.

My beer came and we raised our tankards. 'I'm going to drink to that tackle on the touchline.' He laughed and slapped his thigh. 'My, that was something. The way you just tipped his heels to send him flying, and then that groan, so realistic, and down on your knees beside him. If it hadn't been the home team and at Gartmore, I think the ref might have given the foul the other way. What I can't understand, though: that other time when you backed off, you could just as easy have turned him upside down.'

I showed him the swelling as big as an egg on the back of my head. 'His mother,' I explained briefly.

'His mother,' he echoed, 'well, I never,' and there was a sort of grudging respect in his voice. I felt that it was the judgment of one professional on the work of another when he said, 'She's another one who works well on the blind side.'

'Well, anyway,' I said, 'I expect old Tommy will be back for next week.'

'Oh, sure to,' said the boss. 'Anyway, I got you a game with the Rovers. It's funny, though; last time the Rovers played up here Tommy collected an injury just like yours.' He was laughing again as if at some private joke.

I called for more beer. My head was hurting. I didn't think it was that funny.

31
ᔕ *Waiting* ᔐ

*T*HE summer was well advanced, and still no news of my
bursary. It loomed large in my life. It was my passport to
move on to other things. Out there a big world waited. I was
ready. Would it never come?

The boss and Aunt Kit were only too aware of my unease.

'You're in an awful hurry to leave us,' said Aunt Kit one
day.

I tried to explain to her how far this was from the truth,
how much I should miss them, and the farm, and the horses,
and the work, but that I had no choice, I had to find out about
a wider life, learn more things, see more, travel, expand my
horizons...

I hoped that she might see my point of view. I think the
boss did to some extent. Perhaps he had regrets about missed
opportunity, a move down south perhaps, to try his luck in a
kinder climate. A migration with it all on a train might have
been that once-in-a-lifetime adventure.

There was nothing for it; I had to grit my teeth and wait.
The term would start in October, and I would know before
then.

Meantime, hay was all in the stack. The potatoes were
looking well—the two-row planter had been a boon—and in no
time at all the Irish would be invading.

Sobina had not returned after that eventful year of our first
encounter. She had, I was told, married the loutish suitor and

soon licked him into shape. They had a smallholding now and children, and were making a living. Each year she sent me a big kiss via some dark-haired maiden, which I returned using the same route. I hoped she wasn't confused by the number of colleens who carried the message each year.

Aunt Kit was delighted with her hybrid 303s in the deep litter house. She had never collected eggs so easily before, and as the nest boxes were bedded deep with shavings, she could pack the clean eggs straight into the trays. The lights came on at two in the morning, and Aunt Kit was amazed at the productivity. She put it down to not having cockerels messing about. I had a feeling that might have come from Peg.

The boss had literally looked over the hedge at our neighbour's efforts at silage making. Grudgingly, he admitted that the work next door had looked pretty fair, although spraying that molasses on had seemed a bit messy. I had a feeling silage would feature next year.

A letter arrived from 'Sir' one morning as I was getting the binder ready for harvest. He had seen my application go through. It was being processed; I would hear soon. I was grateful for the letter, but the knowledge that decisions were being made only increased the tension.

Meantime there was a harvest to be tackled. The corn was still standing. Maybe just for once we would have an easy cutting. We heard of combines getting about down south.

'Just think,' said the boss. 'Down south they'll be driving round the fields with the sun burning a hole in their backsides, threshing, no stooking or carting, and balers coming behind wrapping up the straw.' I thought I sensed again a regret that he hadn't put it all on a train; but he had never visited his well-favoured friends. Perhaps he preferred his dream to reality.

We were well into harvest when the news came. My application had been successful, and I was to start my college course in October.

After the initial excitement, a feeling of anti-climax assailed me. I would just about see Fanny's foal before I left. I was going to miss Aunt Kit and the boss, old Top the farm dog, now creaking a bit with age but still doing his rounds, my friends, the Young Farmers' Club, the people on the games circuit. I would miss all the work projects, silage next year, more mechanisation perhaps; and then there were the horses. I was going to miss calling them on the moor early on a summer's morning; working around them, snug in the stable, on a cold winter's day.

I think the boss sensed my mood, because he said, 'Nothing ever stays the same. Things are always changing and you have to keep moving on. You couldn't settle here; you have to make your way in the world.'

The pattern of the boss's life had been determined by the loss of his parents, and the consequent responsibilities of his inheritance; that of mine was fluid. All I was sure of was that I needed to get some formal training, and then I must look for a new job—down south perhaps, or even further afield.

The thought brought a tingle of excitement, but still that feeling of anti-climax, the sense of regret, and the knowledge that I was leaving something solid for an uncertain future, hung around me like a moorland mist.

I had grown roots. I was no longer itinerant; I was established, part of something. Pulling up those roots was going to hurt.

Fanny foaled before I left—a fine colt foal again. A new lad had been hired, but the boss had tactfully arranged for him to start after I had gone.

There was a hint of tears as Aunt Kit hugged me crushingly. 'Come back and see us, lad,' she said. The boss took me to the railway station. My luggage had increased to three suitcases. He didn't wait for the train. His goodbye was gruffly spoken. 'I hope the next wee bugger turns out as well,' he said as he left.

I think he was as upset as I was.

—

I had a few days at home, then started on my new life.

My intention had been to visit the farm regularly, perhaps two or three times a year, but my time seemed to be gobbled up by study and earning money in my spare time.

The Christmas break found me back with my parents and shawing turnips on piece-work for a local farmer. The same farmer engaged my harvest help in the summer holidays. There were few unscheduled hours, what with the need to supplement my grant by free-time working and to have a bit of time for sport or the company of friends.

My studies over, a job offer took me down south and distance complicated matters still further.

Five years had passed before I returned to the farm for a visit.

The boss had moved a little nearer to middle age, but was still eager to explore new ideas and had to hear of all the goings on down south. He was not yet married but was still courting steadily with Peg.

When I quizzed Peg about the long courtship, she said, with a cheery chuckle, 'Oh, he's working round to it in his own time,' and the dig she gave me with her elbow seemed to say they weren't missing out on much.

Aunt Kit had been heard to mention retirement. 'But not just yet,' she told me.

We walked the farm together, the boss and I, looking at tractor-ploughed land, tidy hedges and some winter wheat peeping through. 'After tatties,' the boss explained, 'we just grubbed the soil with the tractor cultivator once the crop was off and drilled the wheat in with the seeder. Behind the tractor, of course,' he added as an afterthought.

In the byre we looked at full udders under two rows of black and white cows.

'You'll be wanting to see the horses,' said the boss. 'I'll just go and see if Aunt Kit's got the kettle on.'

I was grateful for his tact. This had been my headquarters, the centre of my working life, I was glad to be alone with the sight and sounds and smells of horses at rest.

Donald was heading towards old age, but in fine fettle as always. Blossom, well advanced now in years, had retired, her pension consisting of long summer days roaming the moor. Fanny, now in her prime, was a breeder of foals.

The young lad who had replaced me would be a fully fledged ploughman now. The horses, I could see, were in good hands.

I hopped up on to the corn kist and leaned my back against the smooth stones of the whitewashed wall behind. I closed my eyes and let the clicking, clanking, rustling, nostril-blowing sounds, along with the smells of sweet hay, fresh dung and dried sweat, drift over and around me, seeping into my being. The years fell away and I was once again sitting on the horses' dinners en route to the ploughing match, and listening to the sound of horses' hooves on an empty road.

⤚ *Postscript* ⤙

*T*HE small mixed family farm depicted in this book had much to recommend it. It provided a living for a family and an employee, and, not being one-dimensional, it was not extractive as, say, in the production of successive grain crops; fertility was maintained by applications of farmyard manure and grass breaks; crop rotation was adhered to, and although fertilizer was used, nothing was pushed to the limits. However, somewhere beyond the horizon lurked machines which, over the years, would grow into monsters making their own demands; large fields, large farms; the small family farm was doomed. Likewise the power units of the small farm, the work horses, would no longer be needed.

Much of the farm work is touched on in the book, but the system was simple enough. The cows produced the manure to enrich the land and maintain soil structure. The monthly milk cheque took care of running expenses. Poultry and pigs or other sidelines were grist to the mill. Swedes and oats were grown for feed or for sale, and a cash crop of potatoes brought in more income. Sufficient grassland would be allocated for grazing, with some shut up for hay. Land use would be rotated in the traditional manner. A simple, balanced system. All that was needed was a good farmer to make it work. Well, the boss was such a man. His father having died, he took on the responsibility of running the farm at the age of eighteen, and continued until retirement at the age of seventy-six. Over the years the farm grew from 80 acres of good land and 11 acres of moorland to 200 acres. He had turned to sheep in his later years, and for a number of years after retirement he attended markets and bought sheep for other farmers. Some years after my time he met and married Jean. They had a daughter who grew up to be an Olympic hockey player, and as a golfer won the Buchanan Castle ladies golf championship 13 times.

Readers will be pleased to learn that the boss, ninety-five at the time of writing, is well, as indeed is Jean. He plays bowls on a Monday and, as he says, 'I get about, but slower'.

Much of my winter was spent crouching over the handles of a plough, more often than not with badly chapped hands, but I still loved the challenge of the job, holding a dead straight furrow, six or seven inches wide in the lea (grassland) and perhaps eight inches in stubble. Ploughs came in different forms: the Dux, which I used on the farm and at the ploughing match with one wheel under the beam, the two-wheeled sort favoured further south, and on one farm I encountered a two-furrow three-horse plough. I often wondered how far I walked in a day behind the plough. My guess was somewhere between 12 and 15 miles. I suppose it could be calculated; say half an acre a day in six-inch or seven-inch strips or horses walking at two or one-and-a-half miles per hour for eight hours. Perhaps there are just too many variables. I think the words 'the ploughman homeward wends his weary way' just about sum up the end of the day trudge back to the farm, with darkness closing in and the stable work still to be done.

The land was broken down with harrows, and cereal seed was sown using a horse-drawn seed drill. This was hard work on the horses, and we rotated our three horses. Mind you, in an earlier situation I had sown corn broadcast with a two-handed sheet, casting a strip fifteen feet wide. The method sounds a touch biblical, but indeed a lot of fertilizer was applied in this way also.

With the field work involving much manual labour, it was inevitable that peaks in the labour demand—hay time and corn harvest for instance—would occur. I was not aware of any troughs. This need for seasonal help was met by prisoners of war who were housed in Killearn. There was also a Land Girl hostel located in a large house just outside Killearn. We always had the prisoners! I suppose at the time I felt that this was an opportunity missed. However, I didn't entirely miss out, there was the odd invitation to parties at the big house just outside Killearn.

Of course the tractors soon took over and got bigger year by year. Combines appeared, not too big at first, but they soon multiplied and grew. Mechanisation was here to stay. There was little for it but to learn new skills and join the revolution. This I did, working to keep abreast of an increasingly mechanised and scientific use of land.

A fascinating journey, but just now and then I reach for the rose-tinted spectacles. Perhaps the skills learned, and so quickly superseded, were not quite lost. The basics of farming don't change: respect for the needs of the soil, maintaining fertility, the care of livestock, the discipline of working with horses, being part of a small team, dealing with the vagaries of weather. Perhaps those early learned values are still a part of me, and have helped me to enjoy a varied and interesting career in agriculture.

I hope that this journey through a segment of my life will enlighten younger readers as to how it was, and older readers might remember and share with me a time when the cart horse was a daily companion in a hard but satisfying way of life.

Ian Campbell Thomson.